Realizing the Information Future

The Internet and Beyond

NRENAISSANCE Committee

Computer Science and Telecommunications Board

Commission on Physical Sciences,
Mathematics, and Applications

National Research Council

NATIONAL ACADEMY PRESS
Washington, D.C. 1994

NOTICE: The project that is the subject of this report was approved by the Governing Board of the National Research Council, whose members are drawn from the councils of the National Academy of Sciences, the National Academy of Engineering, and the Institute of Medicine. The members of the committee responsible for the report were chosen for their special competences and with regard for appropriate balance.

This report has been reviewed by a group other than the authors according to procedures approved by a Report Review Committee consisting of members of the National Academy of Sciences, the National Academy of Engineering, and the Institute of Medicine.

Support for this project was provided by the National Science Foundation (under Grant No. NCR-9223810). Any opinions, findings, conclusions, or recommendations expressed in this material are those of the authors and do not necessarily reflect the views of the National Science Foundation.

Library of Congress Catalog Card No. 94-65572
International Standard Book Number 0-309-05044-8

Additional copies of this report are available from:

National Academy Press
2101 Constitution Avenue, NW
Box 285
Washington, DC 20055
800-624-6242
202-334-3313 (in the Washington Metropolitan Area)

B-321

Printed in the United States of America

First Printing, May 1994
Second Printing, March 1995

NRENAISSANCE COMMITTEE

LEONARD KLEINROCK, University of California at Los Angeles, *Chair*
CYNTHIA H. BRADDON, McGraw-Hill Publishing Company
DAVID D. CLARK, Massachusetts Institute of Technology
WILLIAM J. EMERY, Colorado Center of Astrodynamics Research, University of Colorado
DAVID J. FARBER, University of Pennsylvania
A.G. FRASER, AT&T Bell Laboratories
RUSSELL D. HENSLEY, Christian Brothers University
LAWRENCE H. LANDWEBER, University of Wisconsin at Madison
ROBERT W. LUCKY, Bell Communications Research
SUSAN K. NUTTER, North Carolina State University
RADIA PERLMAN, Novell Corporation
SUSANNA SCHWEIZER, Digital Equipment Corporation
CONNIE DANNER STOUT, Texas Education Network
CHARLES ELLETT TAYLOR, University of California at Los Angeles
THOMAS W. WEST, California State University

ROBERT E. KAHN, Corporation for National Research Initiatives, *Special Advisor*

Staff

MARJORY S. BLUMENTHAL, Director
LESLIE WADE, Project Assistant

COMMISSION ON PHYSICAL SCIENCES, MATHEMATICS, AND APPLICATIONS

The National Academy of Sciences is a private, nonprofit, self-perpetuating society of distinguished scholars engaged in scientific and engineering research, dedicated to the furtherance of science and technology and to their use for the general welfare. Upon the authority of the charter granted to it by Congress in 1863, the Academy has a mandate that requires it to advise the federal government on scientific and technical matters. Dr. Bruce Alberts is president of the National Academy of Sciences.

The National Academy of Engineering was established in 1964, under the charter of the National Academy of Sciences, as a parallel organization of outstanding engineers. It is autonomous in its administration and in the selection of its members, sharing with the National Academy of Sciences the responsibility for advising the federal government. The National Academy of Engineering also sponsors engineering programs aimed at meeting national needs, encourages education and research, and recognizes the superior achievements of engineers. Dr. Robert M. White is president of the National Academy of Engineering.

The Institute of Medicine was established in 1970 by the National Academy of Sciences to secure the services of eminent members of appropriate professions in the examination of policy matters pertaining to the health of the public. The Institute acts under the responsibility given to the National Academy of Sciences by its congressional charter to be an adviser to the federal government and, upon its own initiative, to identify issues of medical care, research, and education. Dr. Kenneth I. Shine is president of the Institute of Medicine.

The National Research Council was organized by the National Academy of Sciences in 1916 to associate the broad community of science and technology with the Academy's purposes of furthering knowledge and advising the federal government. Functioning in accordance with general policies determined by the Academy, the Council has become the principal operating agency of both the National Academy of Sciences and the National Academy of Engineering in providing services to the government, the public, and the scientific and engineering communities. The Council is administered jointly by both Academies and the Institute of Medicine. Dr. Bruce Alberts and Dr. Robert M. White are chairman and vice chairman, respectively, of the National Research Council.

received a number of inputs from outside itself, most notably several briefings from and consultations with federal officials and representatives of commercial and nonprofit organizations.

The period between the conceptualization and the actual launch of the project in early 1993 saw many changes, most notably a rise in governmental, business, and popular interest in electronic networking and information infrastructure, interest epitomized by the launch of the federal National Information Infrastructure initiative. In the face of these developments, the committee expanded its mission from an examination of issues relating to the NREN program, per se, to an examination of architectural and deployment issues relating to the larger national information infrastructure context in which the NREN program now fits. Consistent with its original charge, the committee paid special attention to the insights, concerns, and needs of the research, education, and library communities. Given the broader focus, however, the recommendations made by the committee are addressed to a broader governmental audience than the National Science Foundation.

NRENAISSANCE is grateful to the many individuals that contributed to its deliberations, including individuals who briefed the committee and others who contributed materials and insights, generally over the Internet. These individuals include Prudence Adler and Ann Okerson, Association of Research Libraries; G. Ernest Anderson, University of Massachusetts; Larry Anderson, Mississippi State University; Eric M. Aupperle, Merit Inc.; William Blumenthal, Kelley Drye & Warren; Hans Bolli, Northern Telecom; Panayot Bontchev, University of Sophia; Laura Breeden, (then FARNET) National Telecommunications and Information Administration; David C. Carver and Karen Sollins, Massachusetts Institute of Technology; John Cavallini and Robert Aiken, Department of Energy; Jill Charboneau, Cornell University; Annette B. Church, California school teacher; Richard Civille, Center for Civic Networking Groups; Robert Collet, CIX and Sprint; Janos Csepai, Budapest University of Economic Sciences; Andrzej Dabrowski, Polish PTT; Bruce Daley, Elaine Wynne Elementary School, Las Vegas, Nevada; Linda Delzeit, California school teacher; D'Ann Douglas, Sallie Curtis Elementary School, Beaumont, Texas; Michael Einhorn, Department of Justice; Robert Gosse and Michael Pollak, Federal Communications Commission; Robert R. Gotwals, Jr., Microelectronics Center of North Carolina; John Gravelle, Merrill Senior High School, Merrill, Wisconsin; Daniel Hartl, Harvard University; Dale Hatfield, Hatfield Associates Inc.; Michael Jeffrey, Nova Scotia Department of Education; Ioan Jurka, Technical University of Timisoara, Romania; Stanley Kabala and Simon Pritikin, AT&T; Thomas Kalil, National Economic Council; Donald Lindberg, HPCC National Coordinating Office and National Library of Medicine; Jack McCue and Howard Palmes,

BellSouth; Steven Metalitz, Information Industries Association; Paul Mockapetris, Information Sciences Institute, University of Southern California; Mark Neibert and Janet Dewar, Comsat Corporation; Michael Nelson, Office of Science and Technology Policy; Roger Noll, Stanford University; Antoni Nowakowski, Technical University of Gdansk; Zoltan Pap, Hungarian Telecommunications Company, Budapest; Paul Evan Peters and Joan Lippincott, Coalition for Networked Information; Gary Ragsdale, Federal Express; Michael Roberts, EDUCOM; David Ruth, Cornell University; Steven Ruth, George Mason University; Anthony Rutkowski, Internet Society; Theodore Schell and Ronald Bracewell, Sprint; Richard Snelling, U.S. Olympic Committee; Thomas Spacek, Bellcore; Steve Stephenson, Waiakee Intermediate School, Hilo, Hawaii; Eric Swanson, John Wiley & Sons; Randy Sweeney, Jordan High School, Los Angeles, California; Frank Withrow, Council of Chief State School Officers; Stephen Wolff and Jane Caviness, National Science Foundation; and Anthony Villasenor, NASA.

It is also extremely grateful to the anonymous reviewers who challenged it to sharpen and focus its arguments. The committee gratefully acknowledges the truly outstanding contributions to this report by Marjory Blumenthal, director of the Computer Science and Telecommunications Board, whose efforts were indispensable to the creation of this report. We also acknowledge the assistance of her staff, notably project assistant Leslie Wade, and of the editor, Susan Maurizi. Responsibility for the report, of course, remains with the committee.

Leonard Kleinrock, *Chair*
NRENAISSANCE Committee

Contents

SUMMARY AND RECOMMENDATIONS **1**

Committee and Its Tasks, 2

The Vision of an Open Data Network, 3

Developing an Open Data Network Architecture, 4

Configuring the Components, 4

Defining NII Compliance and Setting Standards, 6

Factoring in the International Aspect, 7

Deploying the Open Data Network, 7

Research and Education Concerns, 7

Infrastructure Financing: Investments for Research and Education, 8

The Government Role, 10

Long-term Strategy, Management, and Wise Investment, 10

Leadership in Education, 11

Technology Research and Development, 12

Recommendations, 12

The Vision of an Open Data Network, 12

Recommendation 1: Leadership and Guidance, 13

Recommendation 2: Technology Deployment, 14

Recommendation 3: Transitional Support, 14

Recommendation 4: K-12 Education, 15

Recommendation 5: Network Research, 15

1 U.S. NETWORKING: THE PAST IS PROLOGUE **17**

Where We Are Today, 18

Existing Communications Networks and Increasing Focus on Infrastructure, 18

How We Got Here, 22

Today in Transition, 27

Visions of the Information Future: What Might It Be?, 30
The Internet-based Vision, 30
The Entertainment-based Vision, 31
The Clinton-Gore Administration's Vision, 32
Possible Scenarios for Development of a National Information Infrastructure, 32
The Committee's Vision: An Integrated National Information Infrastructure, 34
Converging the Visions of the Future, 35
Technology Impetus, 35
Benefits to the Nation—Last-mile Economics, 36
How Can We Converge the Visions?, 38
Structure and Content of This Report, 38
Notes, 40

2 THE OPEN DATA NETWORK: ACHIEVING THE VISION OF AN INTEGRATED NATIONAL INFORMATION INFRASTRUCTURE 43
The Open Data Network, 44
Criteria for an Open Data Network, 44
Technical, Operational, and Organizational Objectives, 44
Benefits of an Open Data Network, 46
Open Data Network Architecture, 47
An Architectural Proposal in Four Layers, 47
The Centrality of the Bearer Service, 51
Characterizing the Bearer Service, 53
Middleware: A New Set of Network Services, 55
Defining the Higher-level Services, 59
Basic Higher-level Services, 59
More Demanding Higher-level Services, 60
Quality of Service: Options for the ODN Bearer Service, 65
Best-Effort and Reserved Bandwidth Service, 65
Assuring the Service, 66
NII Compliance, 67
Standards, 70
Role of Network Standards, 70
Factors that Complicate Setting Standards, 71
Network Function Has Moved Outside the Network, 71
It Is Hard to Set Standards Without a Recognized Mandate, 72
A Bottom-up Process Cannot Easily Set Long-term Direction, 72

A Top-down Approach No Longer Appears Workable, 73
Commercial Forces May Distort the Standards-Setting Process, 73
Setting Standards for the NII—Planning for Change Is Difficult But Necessary, 73
Issues of Scale in the NII, 74
Addressing and Naming, 74
Mobility as the Computing Paradigm of the Future, 76
Management Systems, 77
Measurement and Monitoring, 77
Security and the Open Data Network, 78
Securing the Network, the Host, and Information, 78
Developing a Security Architecture, 79
Security Objectives and Current Approaches for Reaching Them, 80
Computer System Protection, 81
Protection of Information in the Host, 81
Protection of Information in the Network, 82
Authenticating Users, 83
Control of Authorized Users, 83
Taking a Comprehensive Approach to Ensuring Security, 84
Finding and Balancing Opportunities to Build Toward Convergence, 84
Development of Standards for Television—An Example, 85
Reengineering of the Nation's Access Circuits, 86
Cost and Function in Access Circuits, 87
Options for Incorporating the ODN Bearer Service, 88
Need for Government Action in Balancing Objectives, 89
Acting Now to Realize a Unified NII, 90
Recommendation: Technology Deployment, 91
Research on the NII—Ensuring Necessary Technical Development, 91
Research to Develop Network Architecture, 92
Defining the Bearer Service, 93
Issues for the Lower Levels: Scale, Robustness, and Operations, 94
Addressing and Routing, 94
Quality of Service, 95
New Approaches to Transport Protocols, 96
Network Control Functions, 96
Mobility as the Computing Paradigm of the Future, 97
Management Systems—Monitoring and Control, 98
New Technology for Access Circuits, 98
Middleware and Information Services Support, 99
Navigation and Filtering Tools, 99

Intellectual Property Rights, 100
Computer and Communications Security, 101
Research in the Development of Software, 102
Experimental Network Research, 102
Experimental Research in Middleware and Application Services, 103
Rights Management Testbed, 104
Research to Characterize Effects of Change, 105
Recommendation: Network Research, 105
Notes, 106

3 RESEARCH, EDUCATION, AND LIBRARIES 112
Research, 113
Higher Education, 119
K-12 Education, 122
Lifelong Education, 133
Libraries and the Broadening of Public Interest Networking, 133
Cross-Cutting Observations, 142
Notes, 143

4 PRINCIPLES AND PRACTICE 148
Equitable Access, 149
Flow of Information, 153
Government Information, 154
Privacy, 156
First Amendment, 158
Intellectual Property Protection, 160
Broader Consideration of Ethics, 165
Notes, 166

5 FINANCIAL ISSUES 172
Federal and Other Funding for Networking to Date, 172
Cost of Network Infrastructure, 176
Paying the Price, 183
Imminent Short-term Increases, 186
Recommendation: Transitional Support, 186
Costs of Local Infrastructure and Access to Services, 186
Usage-based Pricing, 189
Flat-fee Pricing, 191
Covering User Charges (Subsidies and Mechanisms), 193
Deriving Specific Funds, 195
Equity, 196
Notes, 198

6 GOVERNMENT ROLES AND OPPORTUNITIES 204
Leadership and Vision, 205
Leadership in Development and Deployment of Infrastructure, 207
Leadership in Education, 209
Recommendation: K-12 Education, 210
Balancing of Interests, 211
Diverse and Fragmented Public and Private Interests, 211
Coordination and Management, 213
Uncertain Technical Expertise, 214
Cross-agency and Uncertain Structure, 215
Recommendation: Leadership and Guidance, 216
Influencing the Shape of the Information Infrastructure, 217
Influence on Architecture and Standards, 218
Influence Through Procurement, 220
Influence on Future Oversight of the Internet, 221
Influence on Network Deployment and Technology Development, 223
Support for Experimental Networks, 224
Approach to Operational Networks and Intermediate Technologies, 224
Research and Development, 226
Conclusion, 228
Notes, 228

APPENDIXES

A FEDERAL NETWORKING: THE PATH TO THE INTERNET 237
B SAMPLE PRINCIPLE SETS 254
C USER SUPPORT SERVICES 262
D STATE AND REGIONAL NETWORKS 265
E INTERNATIONAL ISSUES 269
F KEY TERMS 282

INDEX 287

Realizing the Information Future

The Internet and Beyond

Summary and Recommendations

The potential for realizing a national information networking marketplace that can enrich people's economic, social, and political lives has recently been unlocked through the convergence of three developments:

- The federal government's promotion of the National Information Infrastructure through an administration initiative and supporting congressional actions;
- The runaway growth of the Internet, an electronic network complex developed initially for and by the research community; and
- The recognition by entertainment, telephone, and cable TV companies of the vast commercial potential in a national information infrastructure.

A national information infrastructure (NII) can provide a seamless web of interconnected, interoperable information networks, computers, databases, and consumer electronics that will eventually link homes, workplaces, and public institutions together. It can embrace virtually all modes of information generation, transport, and use. The potential benefits can be glimpsed in the experiences to date of the research and education communities, where access through the Internet to high-speed networks has begun to radically change the way researchers work, educators teach, and students learn.

To a large extent, the NII will be a transformation and extension of today's computing and communications infrastructure (including, for example, the Internet, telephone, cable, cellular, data, and broadcast networks). Trends in each of these component areas are already bringing about a next-generation information infrastructure. Yet the outcome of these trends is far from certain; the nature of the NII that will develop is malleable. Choices will be made in industry and government, beginning with investments in the underlying physical infrastructure. Those choices will affect and be affected by many institutions and segments of society. They will determine the extent and distribution of the commercial and societal rewards to this country for investments in infrastructure-related technology, in which the United States is still currently the world leader.

1994 is a critical juncture in our evolution to a national information infrastructure. Funding arrangements and management responsibilities are being defined (beginning with shifts in NSF funding for the Internet), commercial service providers are playing an increasingly significant role, and nonacademic use of the Internet is growing rapidly. Meeting the challenge of "wiring up" the nation will depend on our ability not only to define the purposes that the NII is intended to serve, but also to ensure that the critical technical issues are considered and that the appropriate enabling physical infrastructure is put in place.

COMMITTEE AND ITS TASKS

To help meet the challenge of building an NII with lasting value and utility, the National Research Council's Computer Science and Telecommunications Board assembled a committee to study issues raised by the shift to a larger, more truly national networking capability. The committee that authored the report brings unique competence derived from members who developed and pioneered the use of the Internet from within the research and education communities and who have researched, built, financed, and operated networks generally.

At the request of the National Science Foundation (NSF), the committee focused on how best to continue to meet the needs of the research and education communities in the midst of the policy and program transition from a National Research and Education Network (NREN) to an NII focus. Drawing on the lessons of the Internet experience, the committee addressed the structural questions of how to build a working NII and considered also the key enabling actions that must be taken by the federal government to achieve an integrated infrastructure that will benefit U.S. society as a whole.

The report that resulted presents a vision of the NII based on an Open Data Network (ODN), whose essential technical components it outlines; points out the potential of the ODN for meeting current and future U.S. networking needs; characterizes the nature of the transition to such an infrastructure; examines how the transition to a larger information infrastructure will affect those communities that have come or will come to depend on the Internet; identifies barriers to and trade-offs in achieving an ODN; and suggests approaches and actions the federal government can take to catalyze and guide the realization of this vision. A main purpose of this report is to express the perspectives of the research and education communities and the vision they generate for an integrated NII.

THE VISION OF AN OPEN DATA NETWORK

There are many possible visions for an NII. Members of the Internet networking communities, for example, look forward to an NII that will continue to provide a laboratory for discovering innovative applications for information technology in research, education, and commerce. Major players in the entertainment, telephone, and cable TV (ETC) sector see movies, games, and home shopping offered over the NII as promising commercial ventures. Motivating the administration's support of the NII are broad social and economic policy considerations basic to improving the quality of life in the United States. Included in the mix of expectations and approaches are the views of various trade, public interest, and professional organizations about the NII's potential for meeting their diverse needs.

The committee's vision of the NII gives form to these diverse expectations as a data network with open and evolvable interfaces. Such a network should be capable of carrying information services of all kinds, from suppliers of all kinds, to customers of all kinds, across network service providers of all kinds, in a seamless accessible fashion. Moreover, the user of an Open Data Network should be able to access this capability as he or she moves from place to place. The network should be scalable in the many dimensions of size, load, services, reach, and utility; should integrate a range of network technology and end-node devices; and should provide a framework for security.

The committee's vision of the NII is based on a 25-year legacy of computer networking in the United States. The current manifestation of that legacy is the worldwide Internet that serves more than 15 million people. Its success is based largely on the Internet's openness, which allows interoperability of all of its attached networks.

Indeed, an Open Data Network includes the following characteristics:

- *Open to users:* It does not force users into closed groups or deny access to any sectors of society, but permits universal connectivity, as does the telephone system.
- *Open to service providers:* It provides an open and accessible envi-

ronment for competing commercial or intellectual interests. For example, it does not preclude competitive access for information providers.

- *Open to network providers:* It makes it possible for any network provider to meet the necessary requirements to attach and become a part of the aggregate of interconnected networks.
- *Open to change:* It permits the introduction of new applications and services over time. It is not limited to only one application, such as TV distribution. It also permits the introduction of new transmission, switching, and control technologies as these become available in the future.

The ability to evolve is a key property of the NII envisioned by the committee. Currently, both the Internet and the telephone network are running out of addresses for their subscribers. In the case of the Internet, a likely outcome is a major change in the protocol suite that will affect millions of computers. Providing the ability to evolve gracefully in any of a number of dimensions is essential to the successful commercialization of the Internet and its integration into a larger NII.

Although the Internet is a clear example of a network with an open architecture, the concept of openness is not universally accepted at present. Witness, for example, the numerous communications networks that deal with closed services (e.g., the cable TV industry, the radio pager industry) or those that have attempted to maintain captive user communities via a proprietary network architecture (usually, users of such networks eventually demand interoperability of heterogeneous equipment). In any case, an open network should certainly allow closed user groups to offer their services on the open network if they so choose.

DEVELOPING AN OPEN DATA NETWORK ARCHITECTURE

Constructing an Open Data Network translates into a number of technical goals and considerations for planning the NII: there is a need for a certain minimum level of physical infrastructure to be provided; for a minimum set of services to be made available; for NII compliance to be defined, to ensure provision of the basic services and to illuminate what is and is not compatible with the open architecture; for supporting standards to be set; for security provisions to be developed; and for oversight and management.

Configuring the Components

Achieving an open network hinges on articulating and maintaining an appropriate architecture. Without a unifying architecture, multiple disparate networks will not only have to replicate common services, but may also implement them in incompatible ways. The Open Data Network proposed in this report involves a four-level layered architecture configured as follows: (1) at the lowest level is an abstract bit-level service, the *bearer service,* which is realized out of the lines, switches, and other elements of networking technology; (2) above this level is the *transport* level, with functionality that transforms the basic bearer service into the proper infrastructure for higher-level applications (as is done in today's Internet by the TCP protocol) and with coding formats to support various kinds of traffic (e.g., voice, video, fax); (3) above the transport level is the *middleware,* with commonly used functions (e.g., file system support, privacy assurance, billing and collection, and network directory services); and (4) at the upper level are the *applications* with which users interact directly. This layered approach with well-defined boundaries permits fair and open competition among providers of all sorts at each of the layers.

In particular, the concept of a distinct bearer service contributes to meeting the key objective of separating the information service provider from the network service provider in order to allow all potential service providers the opportunity to flourish in an ODN environment. To provide for this separation, the committee has structured the protocol stack of its architecture such that it narrows down considerably at the interface to the (open) bearer service layer. Above this narrow "waist" the stack broadens out to include the broad range of options for the transport, middleware, and applications layers. Below this narrow waist, the stack again broadens out to include the many possible technologies for implementing network access, local area networks, metropolitan area networks, and wide area networks. Such an arrangement reinforces the principle of separation and is intended not to prevent the same supplier from acting in two roles, but rather to ensure that individual competitors can enter into the marketplace at either level.

The minimum set of higher-level application services, which builds on the bearer, transport, and middleware services, includes electronic mail, fax, remote login, database browsing, digital object storage, and financial transaction services. As the NII matures, this minimum set should evolve to become more comprehensive. Beyond this minimum set are more demanding services, which include audio and video servers, both broadcast and interactive.

The committee believes it is imperative to develop at the outset a security architecture that will lay the foundation for protections of privacy, security, and intellectual property rights—safeguards that cannot be supplied as effectively on an add-on basis.

Defining NII Compliance and Setting Standards

An NII-compliant network would provide a set of core services implemented in a standard fashion so as to provide for interoperability. *For purposes of this report, the committee defines the concept as follows: An NII-compliant network must provide a technology-independent bearer service and the minimum set of application-level services listed above, that is, electronic mail, fax, remote login, database browsing, digital object storage, and financial transaction services.* If an additional service, such as video, is provided that conforms to NII standards, then such a service will also be said to be, for example, NII video-compliant.

As the concept of NII compliance makes clear, standards are critical to achieving an ODN composed of components and services supplied, owned, and used by a wide range of parties. Establishing standards for an ODN will be a challenge. The networks that will underlie the NII are extremely large, and neither a small group (like the one that shaped early Internet standards) nor the government has either a mandate or the control to set standards.

The forces that bear on the standards-setting process are significant and multidimensional. Among them are competing approaches—the development of a unilateral standard by a dominant vendor versus development in an open, multiparty environment—and the tension between the simple, short-term solution and the longer-term, general and flexible solution. Although in the context of the computer industry standards have evolved through vendors implementing unilateral standards as well as marketplace adoption of ad hoc standards, neither approach necessarily supports development of the standards needed for a sufficiently general and flexible ODN architecture of the scope required for an NII. The short-term profit motive of a commercial provider confronting the uncertain prospects of new ETC service offerings, for example, may not encourage commercial applications that meet open standards. Open interfaces (which can involve some proprietary elements) may increase costs initially or be perceived as offering less competitive advantage than a more proprietary approach. Indeed, there will be considerable pressure to develop closed networks (closed in the sense that only a limited set of services can pass over the network, or only a limited class of users can access the network).

An open architecture does not preclude groups that may choose to operate in a closed mode from doing so. At present, however, the private sector has few incentives to opt on its own for a very general and flexible architecture, and, absent government action, the possibly limited architecture that does emerge, from, say, the ETC community, may be the default architecture for the other constituencies, not least because standards for consumer service interfaces tend to be relatively stable.

Factoring in the International Aspect

A final issue affecting development of an ODN is its international scope, a fact that the mere label "NII" can tend to obscure. Both the Internet and information infrastructure generally are fundamentally international. The international nature of infrastructure will have to be addressed in whatever technical, market, and legal measures are taken to assure smooth communication and interaction between most countries. International connectivity must be maintained and expanded as foreign networks develop and proliferate. Beyond physical access, one or more bodies may be needed to develop and monitor bilateral and multilateral agreements on standards, transborder data flow problems, and transborder legalities generally. In addition, both to assure the maximum usefulness of international connections and to support U.S. vendors, export control restrictions on the sale and deployment of U.S. infrastructure technology should continue to be reviewed and, as appropriate, revised.

DEPLOYING THE OPEN DATA NETWORK

The difficulties of developing an Open Data Network architecture extend to such aspects of deployment as financing the infrastructure, designing and deploying services, regulation, funding of specific uses, determining the breadth and nature of access, establishing features such as protection of privacy, and so on. The number, nature, complexity, and interrelatedness of these issues argue the need for oversight in the development of the NII and raise questions about the appropriate level of government involvement to direct and guide the achievement of a viable ODN.

Research and Education Concerns

Realizing an ODN that benefits the economy and society broadly presents many challenges, several of which are recognized by administration and congressional efforts under the NII aegis. While those efforts address meeting the broadest possible set of needs, the research and education communities present specific sets of needs that should also be recognized and addressed as part of NII planning and policymaking. Financial considerations dominate those concerns, since the research and education communities depend primarily on public funding and generally have very tight budgets.

How the needs of the research and education communities (which are treated here in the aggregate but differ among and within themselves) are addressed will bear on some of the more difficult issues relating to achieving a truly national scale within the NII, including the extent to which access to networks and services can be truly individualized. Choices will have to be made—and their consequences assessed—regarding dimensions that are commonly discussed as principles for infrastructure access and use (equitable access, protection of key rights, and responsibilities). Whereas a goal of the NII is ready access to the network by individuals wherever they may be, beginning with their homes and places of work, the goals of research and educational networking programs have centered on access by an institution (e.g., a school, a library, or a laboratory), which would in turn support access by individuals within the institution.

Infrastructure Financing: Investments for Research and Education

A major concern for the research and education communities is financing access to and use of the NII. To date, access to the Internet has been obtained largely through institutions (such as a university), and those institutions have benefited from federal funding of the NSFNET backbone. Although the imminent elimination of that funding will raise costs far more modestly than many fear (costs will be distributed across user institutions), the change in funding has prompted a more general concern about the possibility of direct charges to individual users. These concerns are magnified among those (notably in the K-12 and smaller higher-education institutions) who have not so far had easy access to the Internet.

User charges are desirable when costs vary with use; in these environments, users are able to assess their investment with respect to competing alternatives. The seemingly "free" nature of Internet use has clearly fostered growth in that use; by the same token, the marginal costs of supplying such common applications as electronic mail have been very low, justifying a negligible price except when congestion is experienced.

Financial burdens are not homogeneous across research and education users, although there has been little systematic analysis of the variation. Researchers with extraordinary demand for network bandwidth are among those who raise questions about the prospect of user charges, and the expected increase in video and audio applications suggests that pricing as a mechanism for rationing access to infrastructure will become more generally needed over time. The committee expects that a variety of pricing schemes will emerge with the broader commercialization of the Internet and its integration with other networks. For purposes of fostering continued experimentation in the evolving infrastructure and to assist in fiscal management in the research and education communities, some flat-rate (subscription) charging alternatives are desirable.

The work of research, education, and other social institutions must often be funded with public monies because it produces substantial spillover benefits for the rest of society that cannot be supported directly by the institutions themselves. Given the government's fundamental interest in the missions of the research, education, and associated library communities, support for information infrastructure access and use is consistent with the overall provision of funding for these communities. However, the allocation of support for information infrastructure calls for a weighing and balancing of competing investment alternatives relating to research and education (e.g., for different kinds of inputs to the processes of research and education). *This committee endorses the singling out of information infrastructure in research and education for targeted support.* The commercialization of the Internet and the integration of applications into research and education suggest that related spending will inevitably be integrated into general research and education budgets, as is the case for telephone and computing costs. In the short term, however, there are a number of transitional issues.

Some researchers and educators who depend on the Internet—for example, individuals with large amounts of use related to their work and those with limited institutional support—may need some assistance in adjusting to increased infrastructure prices. Infrastructure support should be a function of the overall research funding process: agencies underwriting research that requires high bandwidth, for example, should either provide access to the necessary infrastructure (e.g., NSF's new very high speed backbone network service (vBNS) or DOE's ESnet) or expect to cover extraordinary costs—not on the basis of entitlement but as a function of the work being done and its requirements. The critical concerns are avoiding disruption from a sudden and sharp imposition of new user charges, minimizing administrative burdens, and assuring equitable access to public support. Over the long term, more generalized mechanisms may evolve, as discussed in the body of this report.

The power of networking has totally changed the way much of science and technology development is conducted in this country, and a generation of scientists has now emerged who depend on that capability. Yet as regards the benefits of networking, the science research community and portions of the higher-education community have had almost a decade head start on the K-12 schools, smaller higher-education institutions, and public library institutions. It is time to allow these latter communities to conduct the experiment and reap the benefits that upper-tier higher education and research have enjoyed. Movement toward broader deployment and some scheme for ensuring equitable access, with targeted mechanisms such as vouchers, may support that objective.

THE GOVERNMENT ROLE

The federal government has an opportunity to alter, enrich, and extend existing elements of the U.S. communications and information infrastructure and to guide their integration into a more powerful whole. The broadening of focus implicit in moving from a narrower NREN orientation to a broader NII orientation suggests that the federal government can play a variety of roles, each lending itself to expression through a variety of mechanisms. Key roles, which are not mutually exclusive, include:

- Providing leadership and vision,
- Balancing interests and airing competing perspectives, and
- Influencing the shape of the information infrastructure.

Decisions made to meet U.S. needs will bear on international connectivity, which is essential for the NII to fulfill its potential.

Long-term Strategy, Management, and Wise Investment

In promoting the NII, the administration has taken a number of steps that demonstrate the impact that leadership and vision can have. At the same time, sustaining that leadership and establishing the mechanisms needed to implement the vision require additional actions aimed at ensuring both the *development of an architecture* for an Open Data Network and the *deployment* of an infrastructure built from that architecture. The federal government needs to establish and maintain a long-range strategy and program responsible for overseeing the evolution of the NII and its applications and for addressing funding needs over the long term. That strategy and program must incorporate the technical competence needed to develop and deploy the ODN architecture and should, in addition, involve a mechanism for addressing the needs, preferences, options, and constraints of all stakeholders, making sure, in particular, that the transitional and long-term needs of such public-interest communities as research, education, and libraries are addressed.

The need for a program drawing on multiple constituencies is shaped in part by a recognition that the federal role in information infrastructure is inherently limited—the federal government will never be able to invest in infrastructure facilities and services a meaningful fraction of the multibillion-dollar private-sector investment plus the growing investments by state governments in state-based information infrastructure and by all manner of individuals and organizations in local computing, communications, and information access infrastructure. *What the federal government can do is focus its own investments and policymaking to gain the maximum leverage and assure the necessary balancing of interests to make sure that the public interest is met.* As part of that balancing, it will be important that NSF and other agencies not back away from the research and education communities in terms of commitment to their needs for and contributions to information infrastructure.

Despite limited resources for actually building information infrastructure, the federal government has several mechanisms for influencing its shape, in terms of both architecture and deployment. Notable among them are standards- and procurement-related activities, as well as research and development, activities on which this report focuses. Although the federal government cannot set all of the standards required for an ODN architecture, it can (through the National Institute of Standards and Technology and other agencies) participate more effectively in standards-setting processes; it can continue to drive Internet-related standards activities and help to stabilize those aspects of the Internet environment that permit the Internet to function; and it can use its own procurement and information- and service-delivery activities to implement and influence the market for key standards. Of course, it can complement such activities with changes in regulations and tax incentives, options for which are analyzed elsewhere.

Although the NSFNET transition signals a pulling away from the use of procurement to explore infrastructure technologies, the committee notes that there remains value to government procurement of and support for truly experimental networks.

Leadership in Education

There is a wonderful opportunity to use the NII to extend what has been one of the more exciting developments in the Internet, namely, stimulation of curiosity and an increased interest in learning and support for teaching in K-12 education. In those few schools where access to the Internet has been made available, there has been a tremendous influence for the good. Unmotivated students are "turning on"; teachers are networking to share ideas and resources; collaboration among students across the world is taking place; and new modes of learning and of seeking information are developing. Indeed, the use of networking for K-12 education is perhaps one of the hopes we have in this country for repairing what is now accepted as a "broken" educational system. Large though it may be, the cost of providing our youth access to the NII's capabilities must be measured in terms of our increasing need for an educated, highly motivated U.S. work force.

Limited use to date of the current network infrastructure by the K-12 and other education communities indicates a need for more effective governmental leadership. Needed is a specific locus of responsibility and accountability for infrastructure development and use, building on a stronger base of technological competence, within the Department of Education, and the setting of a national agenda for local, state, and federal efforts.

Technology Research and Development

The successes of the NREN program worldwide and the strengths exhibited by the U.S. network infrastructure demonstrate clearly this country's technological leadership in networking and information infrastructure. However, that lead is fast shrinking. It is therefore essential to expand the U.S. research effort into the information infrastructure domain itself, to address issues relating to the representation, transfer, and protection of information as well as to data communications per se. In addition, although network supply and operation will be increasingly commercial, the federal government should maintain its role as a sponsor of both experimental and other, more advanced but still transitional, networks and testbeds. As the history of U.S. network development has shown, sustained support is needed for the timely achievement of true innovations. During the past and current period of government and industrial support and growth, the technology of networking has advanced dramatically. Now, the advance of new technology is accelerating as asynchronous transfer mode (ATM) systems, fiber optics, wireless, mobile access, and other networking developments appear. To maintain this position, the federal government should continue to make funding for research in support of information infrastructure a priority.

RECOMMENDATIONS

The Vision of an Open Data Network

The committee believes that the appropriate future communications infrastructure for the nation will come into existence only if its development is guided by a continuing and overarching vision of its purpose and architecture. Described here in terms of an Open Data Network, this open and extensible infrastructure is characterized by the following technical principles:

- Open to all users,
- Open to all service providers,
- Open to all network providers, and
- Open to change.

RECOMMENDATION 1: Leadership and Guidance

The vision of a national information infrastructure (NII) as articulated by the administration emphasizes significant U.S. social and economic concerns but leaves largely unaddressed a number of critical technical issues. The technical roots associated with the National Research and Education Network program and other components of the larger High Performance Computing and Communications initiative must be effectively and consistently factored into that vision.

The committee recommends that the federal government expand its NII agenda to embrace the Open Data Network (ODN) architecture as a technical framework for the design and deployment of the NII. Required is a stable mechanism to provide the following:

- *—Continued federal leadership in stimulating the development and deployment of an ODN architecture for the NII, integrating the technical, economic, and social considerations basic to achieving a truly national U.S. networking capability.*
- *—Continued federal involvement in the development of standards for the NII. The committee does not conclude that the government should set the standards, but rather that it should support and participate in the ongoing standards-setting processes more effectively, bringing to those processes an advocacy for the public interest and for realization of an open and evolvable NII.*

To this end, the committee further recommends that the federal government designate a body responsible for overseeing the technical and policy aspects of the evolution of the NII and its applications.

The Information Infrastructure Task Force (IITF), which focuses on policy issues, is not sufficient for this role; from the perspective of realizing the ODN architecture, it raises three concerns: (1) by design, the IITF focuses on nontechnical issues and is dominated by nontechnical perspectives; (2) it has the strengths and weaknesses of a cross-agency entity; and (3) it is an evolving construct with an uncertain future. The National Science and Technology Council (NSTC), and its component Committee on Information and Communication R&D, which oversees the High Performance Computing and Communications Information Technology activity, is also not sufficient for this role; it raises these concerns: (1) its mission is to coordinate R&D programs, and (2) it, too, has the strengths and weaknesses of a cross-agency entity.

What appears to be needed is a body that will effectively blend the technical competence of the NSTC with the policy capabilities of the IITF and be able to function for the extended period of time required to develop and deploy an NII with an ODN architecture.

RECOMMENDATION 2: Technology Deployment

The committee recommends that the government work with the relevant industries, in particular the cable and telephone companies, to find suitable economic incentives so that the access circuits (connections to homes, schools, and so on) that will be reconstructed over the coming decade are engineered in ways that support the Open Data Network architecture.

The term "engineering" refers to the process of choosing what equipment to deploy, when to deploy it, and in what configuration to deploy it so that customer needs are met at least cost. The committee concludes that a national infrastructure capturing the ODN architecture will not be widely deployed if competitive forces alone shape the future; deregulation, along likely lines, will not be sufficient to guide the development and deployment of the ODN architecture. While anecdotal, numerous comments from inside the cable and telephone industries suggest that the perceived costs of adding the features that support openness are discouraging the necessary investment in the current competitive climate. The committee therefore concludes that these features will not be incorporated in the evolving national information infrastructure without policy intervention.

Needed now is direct action by government to ensure a planned, coordinated start to deploying the access circuit technology for the NII.

RECOMMENDATION 3: Transitional Support

The committee recommends that temporary subsidies of education and research institutions be considered in cases where the commercialization of the Internet generates exceptional funding distortions.

The last decade has seen a powerful transformation of the activities of higher education and research in those segments where networking has become integral. Side effects of the shift away from a federally funded NSFNET backbone may include temporary disruption and financial hardship for some members of these communities, although overall the impact is expected to be limited.

RECOMMENDATION 4: K-12 Education

The committee concludes that there is a clear and present opportunity to improve K-12 education by the integration of networking into the U.S. educational system. Consistent with recent legislative proposals and the selection of education as one of the emphases in the National Information Infrastructure initiative, the committee recommends the following:

—*The federal government, through the Department of Education, should take a leadership role in articulating to other federal agencies, state departments of education, and other members of the education community the objectives and the benefits of networking in K-12 education. It should define a national agenda that can guide efforts at the state and local level.*
—*Since this leadership requires technical competence, the Department of Education should, in the short term, pursue collaborations with the National Science Foundation and other research agencies, but in the long term should acquire internal technical expertise at a sufficiently senior level.*
—*The Department of Education should set an aggressive agenda for research on telecomputing technology in education. This research should address benefits and applications of high-bandwidth communication and services and the transfer of related technologies to educational applications.*
—*The federal government should continue, and if possible expand, federal funding through matching grants, leveraging state, local, and industrial funds, to stimulate grass-roots deployment of networks in the schools.*

RECOMMENDATION 5: Network Research

The committee recommends that the National Science Foundation, along with the Advanced Research Projects Agency, other Department of Defense research agencies, the Department of Energy, and the National Aeronautics and Space Administration, continue and, in fact, expand a program of research in networks, with attention to emerging issues at the higher levels of an Open Data Network architecture (e.g., applications and information management), in addition to research at the lower levels of the architecture.

The technical issues associated with developing and deploying an NII are far from resolved. Research can contribute to architecture, to new concepts for network services, and to new principles and designs in key areas such as security, scale, heterogeneity, and evolvability. It is important to ensure that this country maintains its clear technical leadership and competitive advantage in information infrastructure and networking.

Networks are emerging as significant tools of social change, creating a new kind of free market for information services and connectivity, and extending the boundaries for research and education in both time and space. The experiences of the research and education communities show that the Internet and NREN experiments are far from over—they have only just begun.

1

U.S. Networking: The Past Is Prologue

The "Information Superhighway" is a constant topic on the evening news. The popular press is awash with comments about a network known as the Internet. Telephone companies, cable TV providers, entertainment conglomerates, and computer and communications hardware and software vendors are forming and dissolving alliances. These are all manifestations of a major transition taking place in our society's communications infrastructure.

A national information infrastructure (NII) now lies within striking distance of becoming a reality. Academia has pioneered the pathways, technology is providing the capability, industry is deploying the networks, and government has primed the funding engine and articulated broad goals. Converging computing and communications technologies have advanced to a point that they can now be deployed and integrated more effectively and economically than in the past; through routine and experimental uses much has been learned about what is required for people to achieve the greatest benefit from these technologies. As a result, electronic communications and information resources are becoming essential to the conduct of business, research, and, increasingly, education, social and government services, and even recreation. Decisions and actions taken today, many driving long-lived investments, will have far-reaching consequences. It is thus imperative that we proceed wisely, drawing on past experience and the knowledge of those who have participated so far in the evolution of this country's capability for network communication.

Perhaps the most useful information to guide us in the current transition can be found in the Internet experience. The Internet has transformed the conduct of research among those who have come to depend on it; it is a holy grail for others, notably the K-12 education community, which should have greater access to it but has not been able to afford that access. Its success, reflected in its dramatic growth and the richness of its uses, is a handsome return on highly leveraged federal investments in the underlying technologies and in access for large segments of the research and education communities for which the Internet was originally built.[1] It both exemplifies and showcases the successful transfer of defense-related technology to the civilian economy. Yet the Internet's success may well be dwarfed by the promised returns from the far larger, more complex, and more integrated NII.

A truly national information infrastructure will be much harder to shape than was the Internet. The Internet arose in a vacuum with little awareness outside the research community, the National Science Foundation (NSF), and other federal research-oriented agencies of what it was and what it could do. Commercial telecommunications and information services, on the other hand, have developed as a result of both market forces and public policy influences ranging from the federal, state, and local regulations that affect telephone and cable television offerings and rates, all the way to the intellectual property protections that affect electronic publication. The broader the conceptualization of the information infrastructure—the farther it extends to embrace information generation and use as well as transport—the greater the planning needed to make it all work together and the broader the relevant policy framework. The scale, scope, and visibility of "wiring up" not only the education and library communities, but also every home and public entity in the United States, present an enormous challenge. Compounding domestic conditions is the fact that whatever measures are taken in the United States must anticipate and sometimes respond to conditions in the foreign networks and infrastructures to which the U.S. infrastructure is and will continue to be interconnected. Ongoing investment aimed at the future NII and the imminent roll-out of infrastructure, together with the uncertainty surrounding the future of the Internet, force consideration now of plans to be made, steps to be taken, and challenging policy issues to be confronted in the development of a broadly useful U.S. information infrastructure.

WHERE WE ARE TODAY

Existing Communications Networks and Increasing Focus on Infrastructure

A network (Box 1.1) is a communication system that connects together geographically distributed users by means of links and switches as

Box 1.1 What Is a Network?

Networks support communication between programs running on groups of two or more computers (hosts; Figure 1.1). Such programs implement applications that may be "running" on behalf of one or more individuals (users). The interchange of data between hosts and users takes place as character (or byte) streams that pass between them; these streams are often segmented into blocks called packets, and these packets flow across the network. One character typically requires 8 bits.

Within the network, there are circuits or links on which communications traffic travels, suitably encoded as electrical signals, light, or radio waves. Circuits may consist of copper cables, fiber-optic strands, or the ether (in the case of microwave, satellite, or radio). The rate at which communications traffic travels through a circuit, or, more precisely, the capacity of the communications circuit, is referred to as the bandwidth of the circuit and is expressed in terms of bits per second. At the ends of circuits are switches or routers that direct traffic from switch to switch on a path from the sending user (sender) to one or more receiving users (receivers). In some applications, such as interactive conferencing, each user may be both a sender and a receiver.

To manage the flow of communications traffic from a sender to a receiver requires control software that generates a set of signals or conventions used by the network components to keep track of the status of the traffic flow. These conventions, or rules, are called protocols, analogous to the meaning of the term "protocol" in diplomacy. Protocols perform many functions. For example, they may specify the characteristics of files to be transferred or provide tests to ensure that traffic reaches the receiver unchanged.

Local area networks (LANs) are networks that operate in a limited geographical area, such as a single building or campus. Common examples are the Ethernet and Token Ring networks. Hosts usually are connected to such LANs. Wide area networks (WANs) connect computers or LANs together over a larger geographical area. WANs aggregate traffic from these sources and are often referred to as backbone networks.

well as control software (i.e., sets of rules that govern the way messages travel among users). A digital communications network communicates information between users in fractions of a second using digital transmission technology. Once information is converted to digital form, it can easily be processed, searched, sorted, enhanced, converted, compressed, encrypted, replicated, transmitted, and so on, in ways that are conveniently matched to today's information processing systems. One of the more significant recent developments in communications is fiber-optics technology, which has totally revolutionized the communications capa-

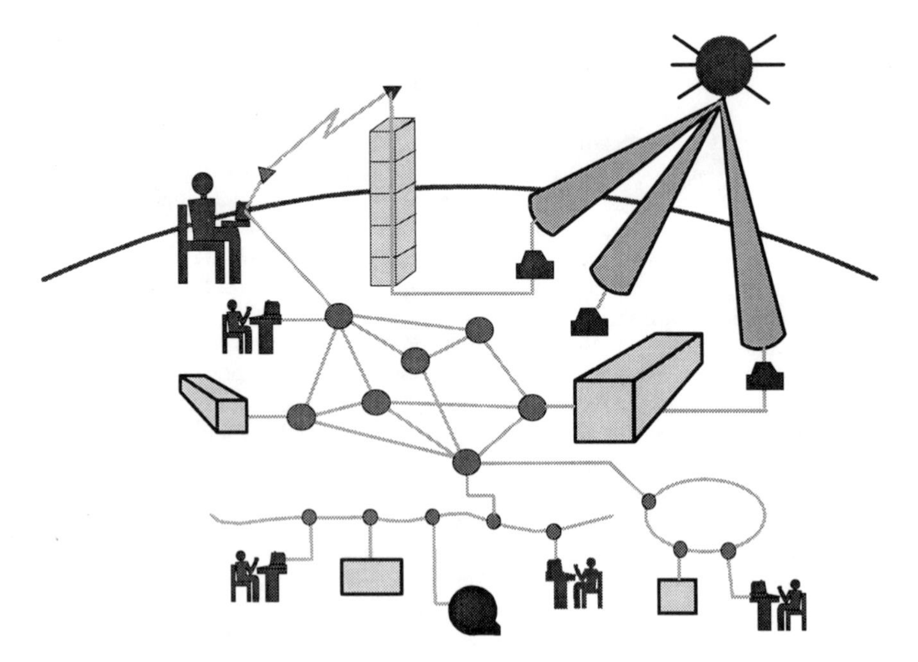

FIGURE 1.1 General network elements. Local or wide area; satellite, land-line, or undersea; fiber, copper, or wireless: all networking components interoperate to provide connectivity among the community of users and host computers.

bility of the underlying links. A second significant development is the design of very fast digital switches based on integrated electronics (i.e., very large scale integrated circuit chips). These, in turn, have energized the development of high-capacity, broadband networks that soon will be capable of transmitting billions of characters per second across our networks.

Today several important networks are being used by society. The telephone network is the most mature component of our communications infrastructure, and the service it currently provides is stable and well understood. A more recent but still quite widespread network is the infrastructure for dissemination of television (and radio) programming, which started with the broadcast networks and has more recently spawned the cable TV industry. The providers of both the telephone network and the cable TV networks (in various partnerships with the entertainment industry) have substantial plans to evolve both the technology of the infrastructure and the range of services provided over those networks.

The Internet (Box 1.2; Appendix A) is emblematic of a newer and

Box 1.2 What Is the Internet?

The Internet is a network of networks. Currently it consists of approximately 20,000 registered networks, some 2 million host computers, and 15 million users. Approximately half the networks are commercial (and this fraction is growing) and half noncommercial; about one-third of the hosts are associated with research or educational institutions. Most of the Internet connections are in the United States, but 149 countries or national entities have connections of one sort or another to international computer networks, with about 63 countries possessing direct connections to the Internet.

The Internet emerged from the packet switching networks developed by the Advanced Research Projects Agency (ARPA) in the late 1960s and early 1970s. It was based originally on the TCP/IP protocol suite but has expanded to accommodate a multi-protocol architecture. In the late 1980s, with the development of the NSFNET by the National Science Foundation (NSF), the available backbone speed of the Internet increased from 56 kbps to 1.5 Mbps and then to 45 Mbps. Backbone network upgrades planned by NSF, the Department of Energy, and the National Aeronautics and Space Administration will introduce even higher speeds, starting at the level of 155 Mbps. With support from ARPA and NSF, gigabit network testbeds have been implemented that are enabling applications that run at end-user speeds in the range of a billion bits per second and more. Operational network capabilities at these speeds likely will follow in a few years.

Each component network is a backbone network that provides connections from hosts and users to the Internet. Host computers are connected to local area networks, some of which are campus-wide, that themselves are connected to metropolitan and/or regional networks. The regional networks are connected to a wide-area backbone network, giving rise to a three-level hierarchy in the United States (Figure 1.2).

different kind of network. The purpose of the Internet, the largest packet switching network in the world, is to provide a very general communication infrastructure targeted not to one application, such as telephony or delivery of TV, but rather to a wide range of computer-based services, such as electronic mail (e-mail), information retrieval, and teleconferencing. Its roots lie in the research community, its original funding came from the Department of Defense, and its current constituency has extended to the commercial world.

In discussions about how to meet national needs for the exchange of information in the coming decade, the focus has now shifted from "net-

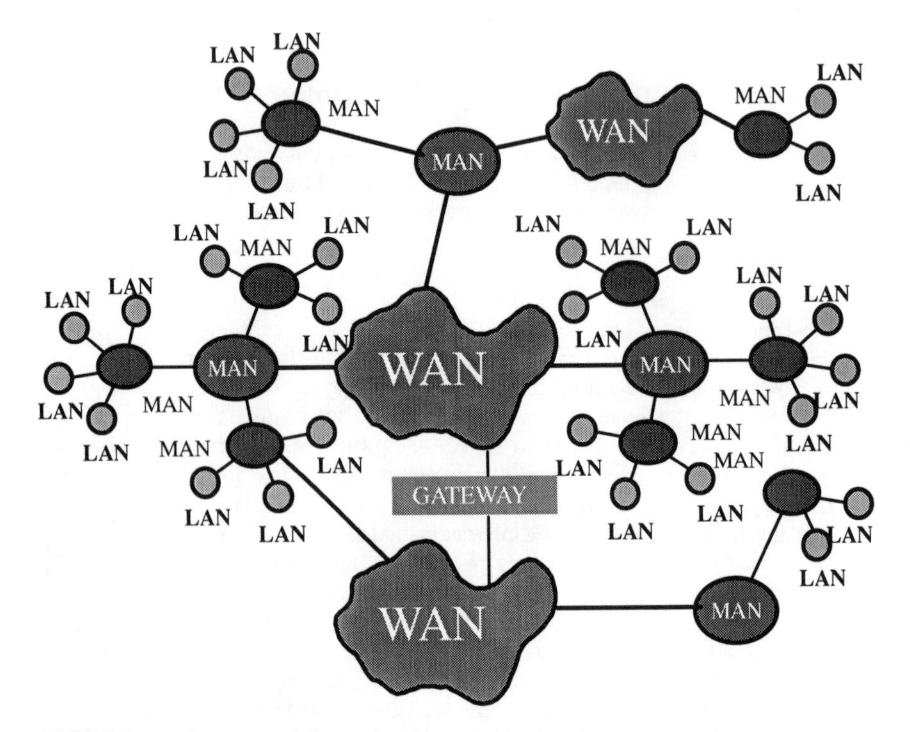

FIGURE 1.2 Internetworking elements. LAN, local area network; MAN, metropolitan area network; WAN, wide area network.

work" to "information infrastructure," which depends on, but involves much more than, networks. An information infrastructure is a framework in which communications networks support higher-level services for human communication and access to information. Such an infrastructure has an architectural aspect—a structure and design—that is manifested in standard interfaces and in standard objects (voice, video, files, e-mail, and so on) transmitted over the interfaces. It specifically moves beyond the simple transfer of unstructured data to encompass issues relevant to higher-level applications. As important as the development of a network architecture is the deployment of that architecture—the implementation of the design and its placement in the field. Innovation and leadership are needed in both.

How We Got Here

The history of the Internet illuminates many issues important to the development of a national information infrastructure. The Internet originated when the first node of the ARPANET (the first packet switching network), sponsored by the Advanced Research Projects Agency (ARPA), was installed at the University of California at Los Angeles in 1969. The evolution of digital networking—first within the computer science community, later as it expanded into different segments of the research and higher-education communities (including research libraries), and more recently with its gradual entry into the K-12 education community and public libraries—has resulted from federal funding decisions that were both farsighted and, at the time, risk-taking. The high-risk, high-payoff approach on the part of government was essential to the acceleration of successful networking (see Appendix A).

When the NSF undertook the expansion of the Internet in 1986, it represented the beginning of a grand experiment as part of the emerging National Research and Education Network (NREN) program. Although the computer science community had become enthusiastic users of the early Internet, the larger science research community had had only limited exposure to the technology, and it was far from clear that it would prove a critical piece of research infrastructure. The NSF, with the support of other federal agencies and the Office of Science and Technology Policy (OSTP), concluded and acted on the wisdom that the correct undertaking was not to perform a small pilot project for the support of science research, but rather to build and operate a network of a size to encompass the vast majority of the nation's research community (Figure 1.3). The result was substantially greater, broader, and faster communication, collaboration, and sharing of new data and insights, plus innovations in the organization, presentation, and retrieval of information by researchers. *For much of the research community, that initial experiment has been a resounding success. It has demonstrated the power of networking to transform a community, to change its operating paradigms, and to build a base of enthusiastic and committed users.*

Over this same period individuals and groups within the education and library communities began to participate in and benefit from access to the Internet. Now increasing numbers of K-12 schools and school systems, community colleges and universities, plus continuing, vocational, and technical education programs across the country are beginning to reach a level of awareness about networking that was achieved by the computer science community 10 years ago. They are initiating a cycle of experimentation, acceptance, and transformation. Meanwhile, demand for access to and capacity on the Internet has mushroomed to professional and personal uses by individuals in all sectors of the economy and society (Figures 1.4 and 1.5).

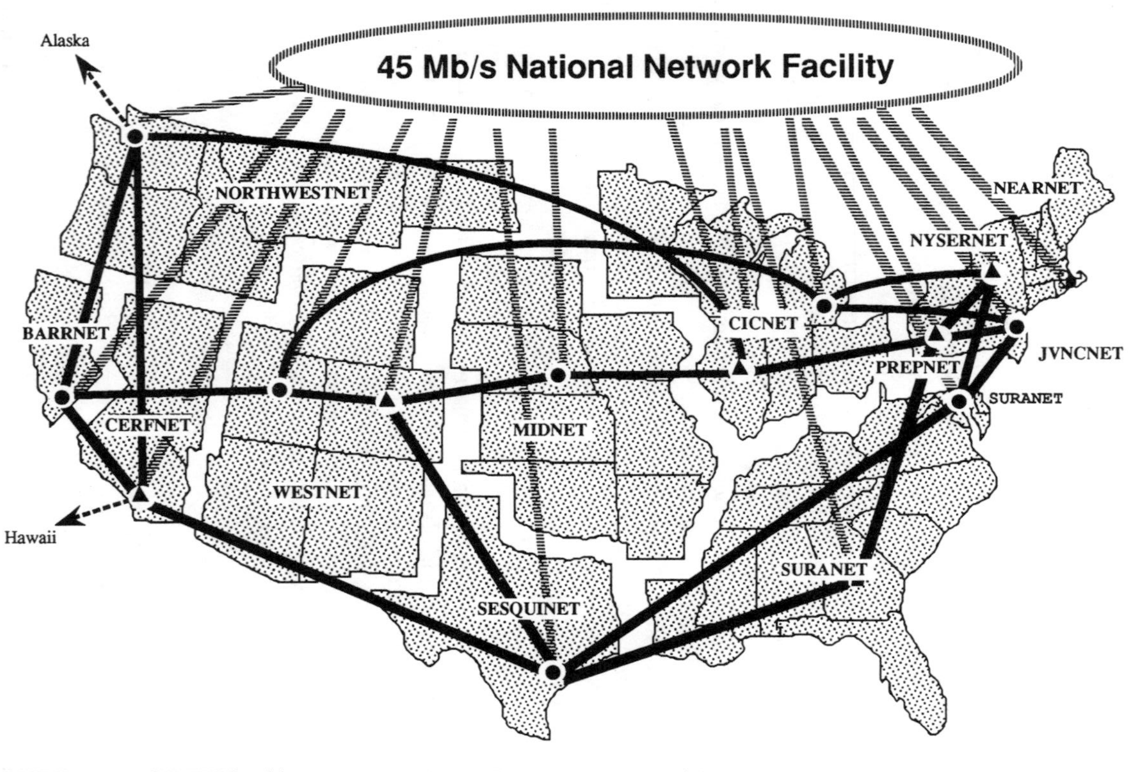

FIGURE 1.3 NSFNET backbone service, 1993. Figure courtesy of the National Science Foundation.

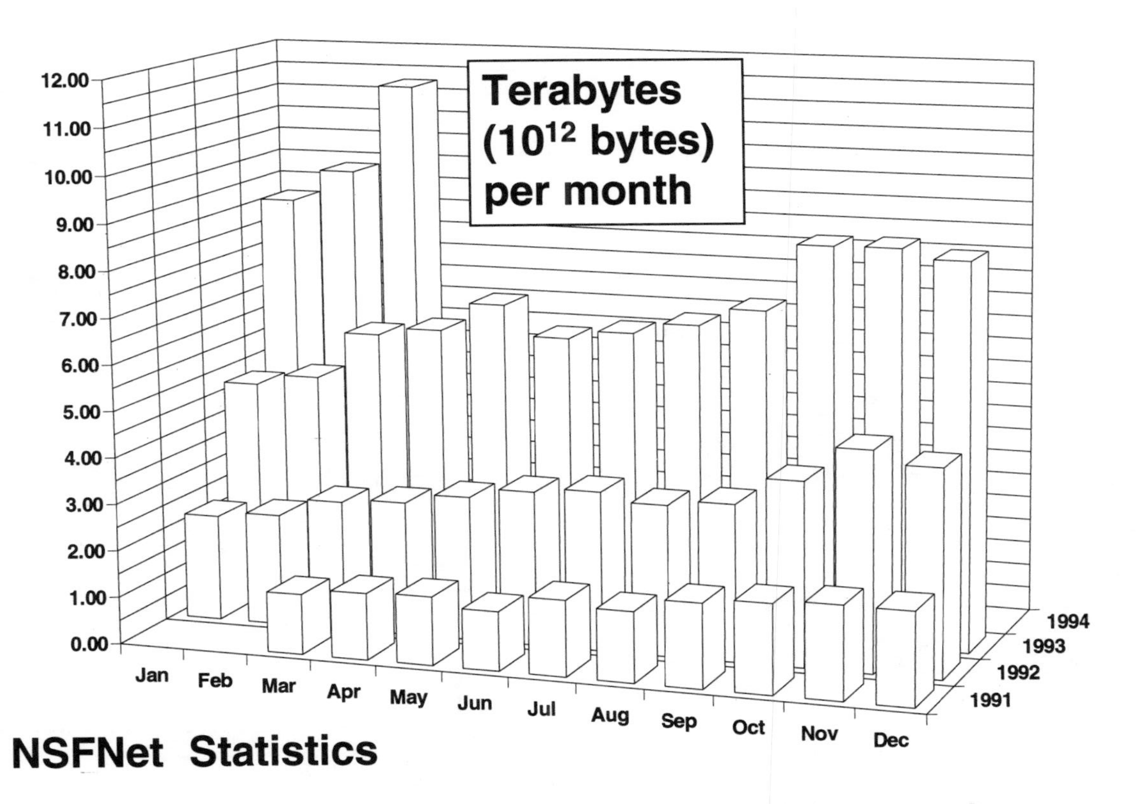

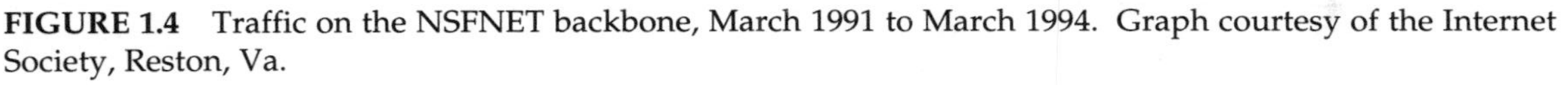

FIGURE 1.4 Traffic on the NSFNET backbone, March 1991 to March 1994. Graph courtesy of the Internet Society, Reston, Va.

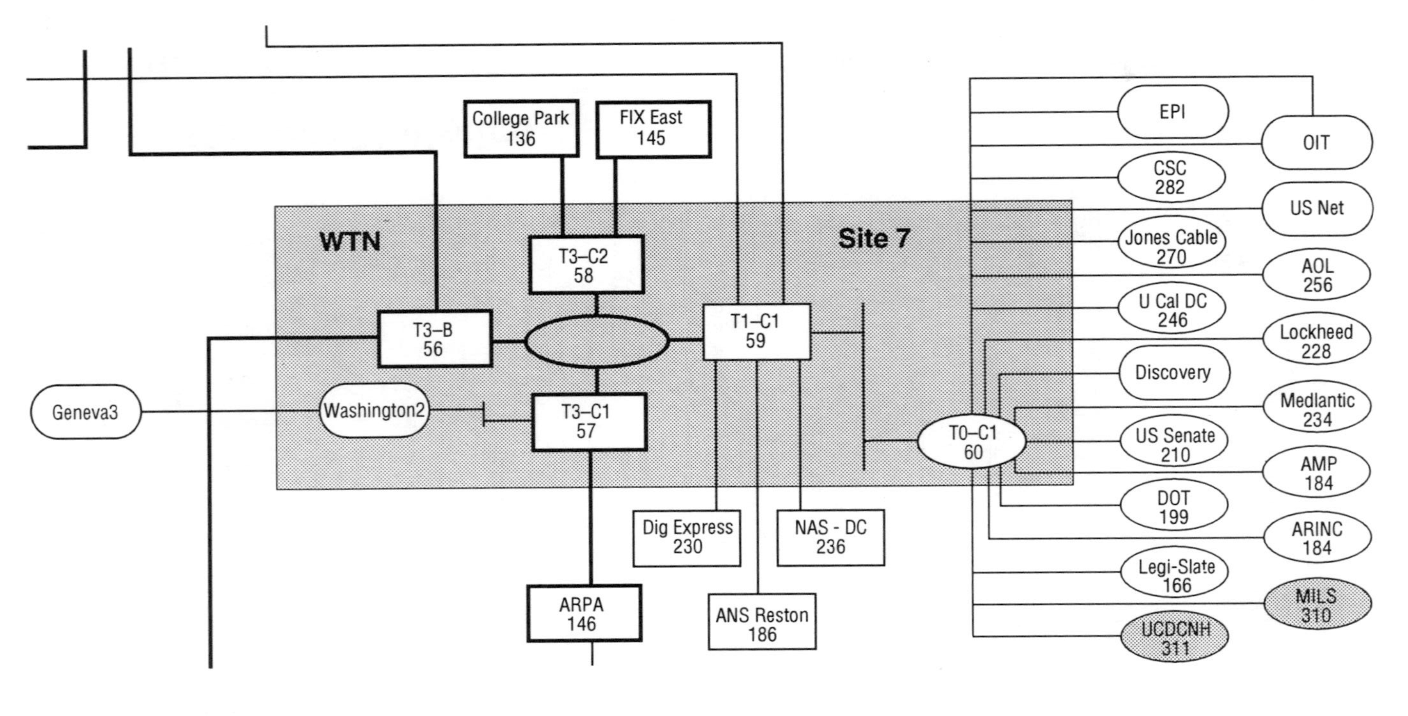

FIGURE 1.5 Internet diversification: Washington, D.C.-area connections to the ANS T3 backbone that underlies NSFNET. Diagram courtesy of Advanced Networks and Services Inc.

Today in Transition

Today the evolution of the U.S. information infrastructure is marked by a rapidly growing and diversifying user population, an increase in private investors, varied information providers, an almost universal set of stakeholders from many different constituencies, and enormous growth of infrastructure applications and services. It is also shaped by a volatile government policy environment with pressures for deregulation of telecommunications providers, conflicting views of the ideal role for government at all levels, and increased political activity by a broad range of stakeholders.

Several of the issues surrounding the future of the Internet and the NREN program were anticipated in the Computer Science and Technology Board's 1988 report *Toward a National Research Network,*[2] which addressed continuing technology needs, the importance of closer interaction with commercial providers of underlying and related services, and the implications of broadening and enlarging the user community. Key changes since the 1988 report include the broadening of attention to education issues and institutions, the growing commercialization of the Internet, the planned withdrawal of NSF from direct provision of Internet backbone service, and the introduction of a more explicit focus on architecture, information resources, and mid- and high-level services.

Another change is that, until very recently, the major transition to gigabit-per-second network technology was discussed as something far down the road and in need of "revolutionary" developments; this was the "phase III" network of the original NREN program. However, the NREN activities launched following the 1988 CSTB report included the fielding and testing of gigabit-per-second technology and applications in a number of testbeds.[3] These testbeds helped to clarify that, indeed, the move to gigabit-per-second speeds could be accomplished in a manageable, evolutionary fashion. Hence, this committee does not warn of a needed revolution, but rather, assumes that the transition can be made smoothly, *if* continued resources are provided for research and development in gigabit technology, architecture, and applications.

At the same time, the committee points out that the history of networking provides cause for concern about how network transitions are managed: ostensibly simple transitions within the Internet environment have proved awkward in practice. For example, as part of "phase II" of the NREN program, the backbone network service was upgraded from T1 (1.5-Mbps) service to T3 (45-Mbps) service, with unpleasant disruptions experienced in network operations. Beginning in 1988, ARPA support for the Internet was reduced, causing confusion among some sets of users and individuals involved in Internet operations. In 1989, the

Box 1.3 The Next-Generation NSFNET

Now that basic network services are readily and economically available commercially, NSFNET will, beginning in 1994, evolve into a far smaller but very high speed national backbone network for research applications requiring high bandwidth (Figure 1.6). In a recent solicitation, the National Science Foundation (NSF) requested proposals to:

- Establish an unspecified number of network access points (NAPs) where regional and other service providers (route servers) will be able to exchange traffic and routing information;
- Establish a routing arbiter to ensure coherent and consistent routing of traffic among NAP participants;
- Establish a very high speed backbone network service (vBNS) linking the NSF-supported supercomputer centers; and
- Allow existing or realigned regional networks to connect to NAPs or network service providers, which will connect to NAPs, for interregional connectivity.

The NAPs will provide connectivity to mid-level or regional networks serving both commercial and research and education customers and will also provide access to the vBNS.

With respect to regional networks, the solicitation addressed only interregional connectivity. Ongoing complementary intraregional support will continue and will be funded at constant or rising levels. These efforts include the Connections program, which provides grants either to individual institutions or to more effective or more economical aggregates. A separate effort is anticipated to address intraregional connection of high-bandwidth users to the vBNS.

Interconnecting the NSF supercomputer centers, the vBNS will be part of the Internet. It is expected that the vBNS will run at a minimum speed of 155 Mbps and that low-speed connections to NAPs will be routed elsewhere.

SOURCE: Adapted from Committee on Physical, Mathematical, and Engineering Sciences, Federal Coordinating Council for Science, Engineering, and Technology, Office of Science and Technology Policy. 1994. *High Performance Computing and Communications: Toward a National Information Infrastructure.* Office of Science and Technology Policy, Washington, D.C., p. 37.

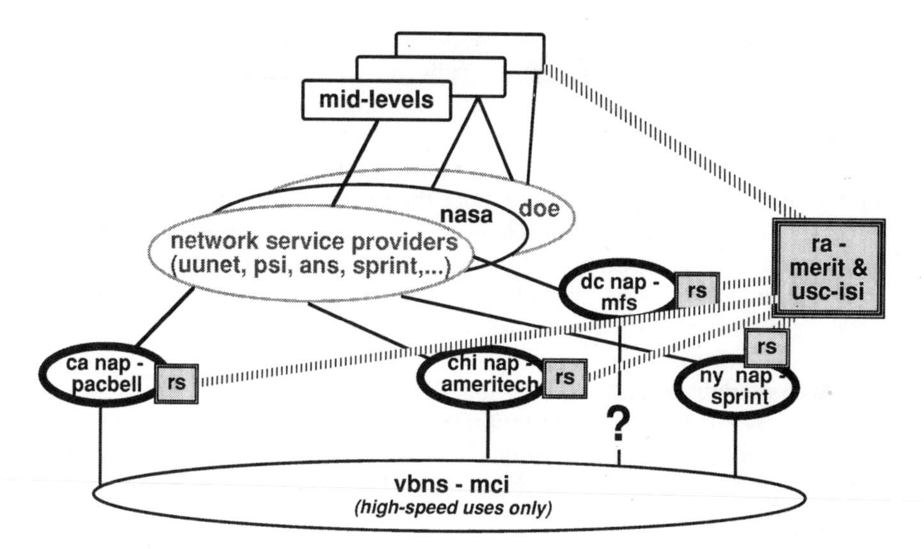

FIGURE 1.6 Elements of the next-generation NSFNET called for in the solicitation outlined in Box 1.3. NAP, network access point; ra, routing arbiter; rs, route server; vBNS, very high speed backbone network service. Diagram courtesy of the National Science Foundation.

ARPANET was decommissioned, in some cases with abrupt termination of circuits and connectivity. These experiences provoke anxiety within the research and education network communities about any kind of transition, beginning with the currently contemplated shifts in NSFNET (Box 1.3). The nature and pacing of a variety of transitions have become real and pressing issues; getting there is at least as important as the nature of the destination.

The concept of the NREN and its research and education applications, which were built on the premise of government-funded special-purpose networks, must now be integrated into a construct built on commercial networks offered in an emerging competitive marketplace. This is part of a process of maturation, which will have many dimensions, including the need for more professional management. The need for that integration is reflected in the recent evolution of the NREN program. In particular, the NREN program has been complemented by a new (fifth) component of the High Performance Computing and Communications (HPCC) initiative, the Information Infrastructure Technology and Applications program, and the entire HPCC initiative has been made a component of the larger National Information Infrastructure initiative.[4]

In view of the shift to a larger, more truly national NII context, it is necessary to determine how to meet underlying needs of the research, education, and library communities reflected in the NREN program. The federal government has a historic and ongoing interest in the conduct of research, and the Internet experience makes clear that networking and information infrastructure are inextricably tied to the future of research in this country. At the same time, both the recent NREN program experience and its broadening clientele show that achieving widespread and economical access to information infrastructure is difficult. Moreover, the problem of access is much larger, in terms of the number of needed access points and participants, for the educational community than for the research community.

A challenge, then, is to develop the leadership and the champions that will extend the implementation of networking to the education and public library communities, building on and linking to the NSF-driven experiment for the research community. The expansion of information infrastructure into these communities is the beginning of a broader expansion into the home implicit in an NII.

VISIONS OF THE INFORMATION FUTURE: WHAT MIGHT IT BE?

The Internet-based Vision

One vision of tomorrow's networking derives from the experience with the Internet. As an open network it has from the outset been a laboratory for discovering innovative ways to use information technology.[5] For example, through its distributed and broad-based membership, the Internet has given rise to a phenomenon in which services of all kinds spring up suddenly on the network without anyone directing or managing their development. These services are offered free to anyone who asks. Indeed, more generally there is any-to-any connectivity of all kinds. For example, "newsgroups" abound on the Internet, covering thousands of subjects of interest to its members; these range from topics on all aspects of computers and research to hobbies, travel, and personal advice, among many others. These newsgroups embody a new kind of social interaction that no one had predicted—just as the value of electronic mail had not been predicted when the ARPANET was conceived. Further popularizing the Internet is the recent introduction and runaway growth of the Mosaic service, which allows delivery of pictures and graphics to users' screens from a variety of servers on the Internet. Hypertext documents can now contain high-quality graphics and even moving picture clips, and through "hot links" users can navigate through a document in a highly customized fashion. This capability has the potential to change significantly the way we use stored knowledge.

Such spontaneous generation of unforeseen yet enormously popular services—which is encouraged by the Internet as a distributed information and communications system—is a constant source of pleasant surprise today and heralds future potential as we move into an era of truly interactive information via the NII. The success of the Internet is such that commercial carriers and other private investors are not only offering but also demanding to be allowed to develop and manage the infrastructure and to provide information services to be delivered over that infrastructure.[6]

The Entertainment-based Vision

Another vision of the information future derives from the commercial opportunities that have been identified for new services in the entertainment sector. It grows out of the video delivery services provided by the cable TV industry and is receiving great attention. Although a considerable market is contemplated for infrastructure serving business needs,[7] the entertainment sector is concerned primarily with residential consumers and therefore is popularly linked to the notion of universality. Whereas the Internet today serves approximately 2 million host computers, the entertainment, telephone, and cable TV (ETC) networks will likely serve 50 million to 100 million, and those computers will, in many cases, look like cable TV set-top boxes. The ETC communities are to some extent joining forces; cable TV networks will in the future be interactive, and telephone systems will have greater bandwidth in order to accommodate video.

The ETC push reflects an industry agenda that is substantially independent of the Internet, of its users (the research, education, and library communities), and of its directions. Almost certainly it is from the ETC industry complex that the major financial resources will come for the NII as we wire up to serve the country: it plans to infuse billions of dollars into the required infrastructure. The major players who are currently trying to capture this market see movies, games, and home shopping as very large business and investment opportunities.[8]

Recently, uncertainties about market demand and payback have caused many executives and financial analysts to moderate their expectations and plans. Several business deals have collapsed, signaling both difficulty and delay, which have been attributed to technical problems, lack of consensus on market trends, and regulatory pressures or uncertainty.[9] Morgan Stanley economist Stephen Roach, for example, questions the potential for growth in demand, estimating that U.S. households currently spend about $160 billion on all forms of "multimedia" (including TVs, video cassette recorders, and video tapes; computers; audio equipment; TV and audio repair; telecommunication links; and applications such as cable TV, video cassette rentals, and movie admissions), more than double the 1983 level of $74.4 billion. This spending level is about 3.3 percent of total disposable personal income and may account for 10 percent of all U.S. household discretionary outlays.[10] Differences in corporate culture, outlook, and experience further cloud the nature and timing of ETC interactions, raising questions about how much and what kind of private investment in information infrastructure will be made and when.[11]

The Clinton-Gore Administration's Vision

One barometer of the political forces that will help to shape a national information infrastructure is the set of efforts by the Clinton-Gore administration to frame the discussion by offering its own NII vision (Box 1.4).[12] The administration envisions an encompassing information infrastructure that integrates and rationalizes various ongoing network developments. Motivated more by social and economic policy considerations than by technology advancement, the administration's view emphasizes access by individuals regardless of means or location and use by industries and public institutions in support of national goals ranging from competitiveness to delivery of government services. Thus, for example, it combines goals of universal access to broadband, interactive services with goals of connecting all classrooms, libraries, hospitals, and clinics by the year 2000.

Possible Scenarios for Development of a National Information Infrastructure

It is clear that the emerging NII will be a hybrid, rather than a linear descendent, of the earlier technologies. What is not clear is how prevailing forces will interact and what the nature of the resulting infrastructure will be. It is difficult to predict the outcome of something that is changing so rapidly and seems to generate so many misconceptions as it evolves.

One possible consequence of the diverse expectations for an NII is that many disjoint network technologies will be deployed, each dedicated to a particular networking objective and each providing its own set of restricted services. Another outcome is that the competencies and interests of the academic and commercial communities may come together and drive a concerted effort to create a network that can be the basis for

Box 1.4 Selected Tenets of the Administration's NII Vision

- Encourage private investment.
- Provide for and protect competition.
- Provide open access to the network.
- Avoid creating a society of information "haves" and "have nots."
- Encourage flexible and responsive governmental action.
- Protect privacy and copyright.
- Ensure that the United States remains a leader of the information age.
- Encourage the emergence of a new kind of communications service provider that offers switched, broadband digital transmission services to the home and office.
- Provide for interoperability.
- Create new jobs, new companies, new technologies, new products, new services, new markets, and new business opportunities.
- Improve delivery of health care.
- Lower prices.
- Create expanded diversity of choice for U.S. consumers.
- Bring into millions of homes information that will enrich people's economic, social, and political lives.
- Spur economic growth and increase competitiveness.
- Democratize information, giving all Americans access to the information they want and need, where and when they want it, for an affordable price.
- Provide educational opportunities created by long-distance learning and networks of schools and universities.
- Provide community empowerment that comes from linking citizens with their governments.

SOURCES: Office of the Vice President, the White House. 1994. "Vice President Proposes National Telecommunications Reform: Bring the Information Revolution to Every Classroom, Hospital, and Library in the Nation by the End of the Century," press release, January 11, e-mail version; Office of the Vice President, the White House. 1994. "Background on the Administration's Telecommunications Policy Reform Initiative," press release, January 11, e-mail version; Remarks Prepared for Delivery by Vice President Al Gore. 1994. Royce Hall, UCLA, Los Angeles, Calif., January 11, fax; "Administration White Paper on Communications Reforms," January 27, 1994, e-mail version.

commerce, education, and research but that is still insulated from the infrastructure of the ETC industry complex. A third alternative entails a fuller integration, with the interests of the ETC industries, the commercial communities, the education and research communities, and the general public all being served by a common infrastructure (Figure 1.7).

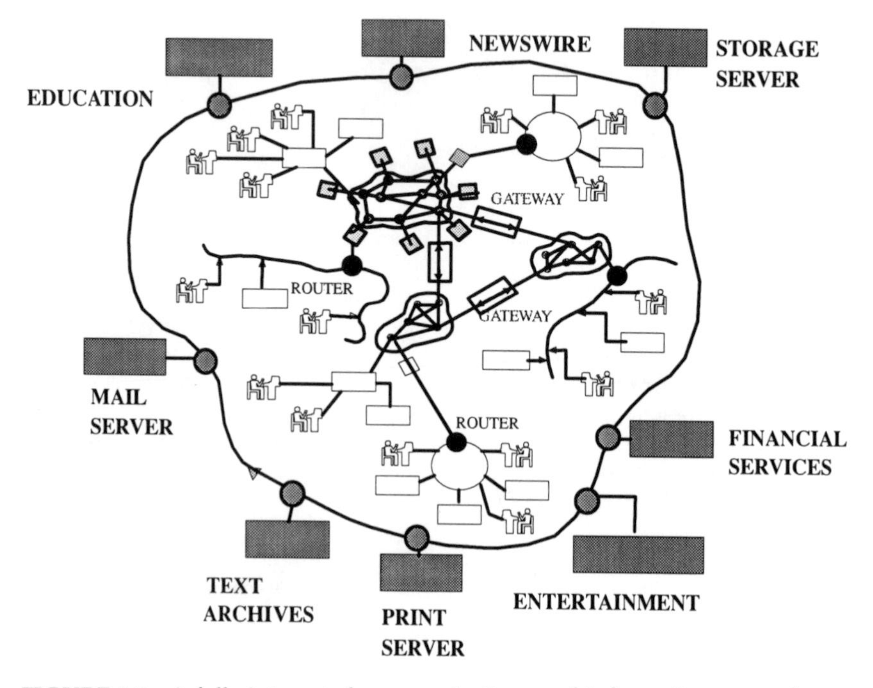

FIGURE 1.7 A fully integrated communications and information network supporting interoperability and a range of services, indicated by examples.

The Committee's Vision: An Integrated National Information Infrastructure

Full integration of communications and information infrastructure (the third alternative above) is supported by the committee, which believes that integration should be an explicit goal if the networks of today are to evolve to a general information infrastructure serving society with a broad range of services. The committee, in developing its vision of a future NII, has been strongly influenced by the Internet's openness, a characteristic that has been key to its unprecedented success. It therefore characterizes its vision in terms of an Open Data Network (ODN).

A national information infrastructure should be capable of carrying information services of all kinds, from suppliers of all kinds, to customers of all kinds, across network service providers of all kinds, in a seamless accessible fashion. The long-range goal is to provide the capability of universal access to universal service, but one that goes beyond a lowest-common-denominator approach. Much of this report is shaped by the committee's belief in the importance of defining and achieving such an overarching vision for tomorrow.

CONVERGING THE VISIONS OF THE FUTURE

Although there are valid historical reasons that the entertainment- and information (i.e., Internet)-based visions have not yet been harmonized, the committee believes it is in the national interest to integrate these visions into a more comprehensive model—one that will support diverse objectives and help ensure national well-being on many fronts. The Internet, the commercial networks, the drive by the ETC complex, the consumer marketplace, and the needs of the research, education and library communities are all interrelated; the infrastructure, deployment, and services of the NII can and should all be highly synergistic.

Technology Impetus

Historically, the technical features of the networks underlying the Internet and the networks for TV distribution have been very different, but broadband digital transmission and fast, inexpensive computing are diminishing some of the need for distinctive technological solutions. Past networks for television distribution have been characterized by specialization to that one single, high-bandwidth data type and a lack of switching inside the network. Standards for television transmission over the air and over cable have been based on analog rather than digital transmission. The standard for analog transmission of television, NTSC (after the National Television Systems Committee), is now embedded in every television receiver, and these television sets are expected to receive 30 or more frequency-multiplexed NTSC channels simultaneously and to select from these the one that is to be displayed on the TV screen. Thus, currently, it is the TV set itself, not the broadcast or cable TV network, that does the switching.

The Internet operates very differently. From the beginning, the Internet was designed to carry a range of higher-level services, not just a single application such as NTSC video. The information to be carried was digitally encoded, and any digital data could be transported. Instead of broadcasting all data sources to all destinations on the network, the Internet uses internal switching to direct data only where it is needed. However, the high cost of long-distance circuits meant that the bandwidth of Internet circuits was much lower than that of television networks; this limited transmission capacity could not support real-time transmission of video.

That situation is now changing, and many, if not most, information networks will soon have the capability to transport and switch video in digital form along with other forms of digital information. The Internet has increased in speed to the point that small quantities of audio and low-quality video are now transported over its infrastructure. At the same time, the cable networks are evolving to support a more general range of services, not just broadcast of NTSC video. In some cases, the cable service providers are planning to carry digital signals, a circumstance that raises the question of interoperability among the various networks that transport electronically encoded information. *The distinction among the infrastructure needs of these various services is rapidly disappearing; thus it is natural to consider an architecture that can, indeed, support interoperability among these networks.*

Benefits to the Nation—Last-mile Economics

At least three important benefits will accrue to the nation if different technologies for providing video, audio (including telephony), and data communications services are interoperable, are linked by recognized interfaces and standards, and are capable of sharing a common infrastructure when appropriate. These benefits, which will accrue because economics will probably dictate the use of different technologies in different service situations, are the following:

- A more coherent, forward-looking, and versatile reconstruction of the circuits that connect homes, offices, and schools to the wide area networks (the "last-mile" circuits);
- Economy and convenience to the end user resulting from having all forms of electronic information accessible through known, interoperable formats. Information services in the future will incorporate combinations of audio, video, and computer data. It would be best if all three could be delivered in a common framework with common conventions for representation and interfacing; and
- Full connectivity between any two appliances attached to the network. Already available are personal computers that support audio communication and display television signals. Whether information is distributed inside the home through one or several interfaces, the devices themselves, such as televisions, computers, consumer electronics, or telephones, can, if they have common signaling conventions and display formats, become a part of an integrated end-user networked environment.

Most U.S. homes have a telephone line that connects them to the local telephone company's network, and in the street outside 90 percent of these homes there is a cable TV line. This connection, called the "last mile" in the communications industry, represents a tremendous investment that cannot be easily replaced. We will be able to reconstruct the last mile only once in the next decade or two (even though we may get two or three such links into the home: one from the telecommunications industry, one from the cable TV industry, and one from the electric power utility industry). As noted above, in the vast majority of cases these in-place facilities provide analog, not digital, service and do not provide the capabilities to support either the emerging entertainment sector or the advanced services being developed for the Internet, much less the NII. The telephone link, with current analog modems, can typically support speeds of 9,600 or 14,400 bits per second (bps), or if converted to a digital service, 128,000 bps (the "2B + D" narrowband integrated services digital network service). These are not adequate rates for quality television. The cable TV line, while it has more raw bandwidth, does not permit any delivery other than NTSC video and does not in most cases provide a reverse channel for traffic leaving the home.

If existing telephone technology were used to provide a service with a bandwidth sufficient for advanced information services, the costs with today's price structure would prove intolerable. At current commercial rates, the average fee for attachment to the Internet at 56 kbps is about $15,000 per year. This rate will come down as usage grows and technology improvements reduce the cost, but $15,000 per year puts advanced Internet use far out of the reach of most home owners.

Unless the price for access is reduced so that it is much more in line with the price of telephone service, NII services to the home will not be widespread. Cost reduction requires investment in new equipment for the access network (such as a multiplexer on the street corner and a shared fiber connection to the central office, as required to support the subscriber loop carrier systems being installed in local exchange carrier access networks), and this, in turn, depends on sufficient, demonstrated demand for the service.

Telephone and cable TV companies are today preparing to make substantial upgrades to their equipment. Trials of new systems, intended in part to gather market research and therefore not necessarily indicative of the costs of full commercial deployment, are being conducted in Florida, California, New Jersey, and many other states. Directed toward the services of the entertainment sector, these trials include interactive television and video on demand. Many billions of dollars will likely be spent on reconstructing cable TV and telephone lines before the end of this decade. That is, old plant will be updated, either because it is now providing poor-quality service or because it cannot do what customers want it to. Cost reduction will require the development and installation of new communications equipment in central offices and other premises, if not the laying of new cable or fiber at subscriber locations. Justifying this investment will require sufficient demonstrated demand for the resulting service. To ensure construction of a broadly useful NII, the upgrade plans now being developed should take into account the needs and services of both the entertainment and information sectors.

Consideration of the last-mile economics is the most important of the issues that impel a unified vision of the NII. However, two other points are also worth considering. One is that the boundary between entertainment and information is not absolute. Video is becoming a part of stored information, and entertainment, as in interactive television and multiplayer games, depends on the exchange of control information among participants. If there were a uniform coding of video used for entertainment and information services, then the commonality could easily be recognized and integration more easily focused.

The other is that today the television and the computer cannot interact or exchange any sort of data. Both have displays, and both attempt to provide visual information to the user, but there is no interplay between them. This lack of compatibility is an issue in terms of cost to the consumer, and in loss of flexibility to take advantage of new services.[13] As we bring into the home a next generation of networked devices, this problem will become more generally relevant: new devices for telephony, for entertainment, and for information access should all interwork.

How Can We Converge the Visions?

How can the various distinctive visions of a national information infrastructure be integrated to support interoperability and a common NII objective? All have a large constituency and significant fixed investments in their installed technology base. Until recently, technical, economic, and regulatory barriers have yielded no reason to strive for interoperability, and standards have been developed independently as well. It will be necessary to find specific points at which the convergence can be encouraged, to take account of the real economic issues, especially in the entertainment sector, and to take an active role in defining and fostering the overarching vision.

STRUCTURE AND CONTENT OF THIS REPORT

This report has three essential components. The first is the committee's presentation of a detailed framework for an integrated information infrastructure, which it calls the Open Data Network. Chapter 2 explores the nature of the architecture essential to achieving true open information networking, discusses the key actions needed to implement that architecture, and points out how achieving the goals advanced by various stakeholders, providing the desired benefits, and controlling costs will depend on continuing research and development.

The report's second and third components concern deployment of an ODN architecture, although it is difficult to totally separate development of the architecture from its deployment. Chapters 3 through 5, the second component, address the planning needed to provide connectivity both within and between the research, education, and library communities in the larger national context and within the evolving framework of the NII.

In Chapter 3 the committee examines the history of the research, education, and library communities' use of information infrastructure, assessing difficulties, benefits, and prospects in order to provide insights into the larger challenge of deploying a true NII and to express the continuing needs of communities that have so much to contribute to and gain from an NII. Generating and sharing information are fundamental to the activities and missions of these groups, which have innovated many of the information infrastructure elements now becoming more widely accessible. As a broader range of users develops, these three communities will shift from the majority position they hold today to a minority force in the evolution of national network-related initiatives. The challenge is how to balance and harmonize the varying needs and desires of the existing research, education, and library communities with those of the emerging communities of users while keeping in focus the goals of serving the educational, professional, and social needs of society at large.

Chapter 4 discusses the experiences and perspectives of these communities as they relate to commonly advanced principles—such as equitable access, protection of key rights, and responsibilities—for the design and operation of an NII. Chapter 5 addresses the challenges confronted by the research, education, and library communities in financing access to and use of a commercially provided information infrastructure.

Representing the third component of this report, Chapter 6 considers emerging roles for the federal government, which has an opportunity to help guide the expansion of the existing networking infrastructure in ways that will achieve an Open Data Network architecture and an appropriate balancing of interests in deploying such an architecture. There are several options for and issues affecting how the federal government can act on that opportunity and how effective the result will be.

Several appendixes provide complementary information. Appendix A outlines the federal government's history of involvement in networking as it relates to the NREN program and the Internet, focusing on NSF's role and describing in some detail the Internet's evolution and how it works today. Appendix B provides illustrative sets of NII principles advanced by a variety of groups. Appendix C characterizes user support needs. Appendix D outlines state and regional networking activities.

Appendix E assesses international dimensions of the Internet experience and their ramifications for the international connections that are inevitable for an NII. Appendix F defines key acronyms and terms.

The committee's several recommendations, which are reproduced in the summary, are presented in a different order in the body of the report in the context of the issues that occasion them.

Readers should keep in mind a number of distinctions made in this report. One is the distinction between the research, education, and library communities as opposed to the commercial and business communities. Another is the distinction between the Internet and commercial networking communities and the entertainment, telephone, and cable TV communities. A further breakdown within the research, education, and library communities distinguishes the research segments (i.e., university-level research in nonscience disciplines, science research, and the work of research libraries) from the nonresearch segments (i.e., K-12 education, most higher and continuing education, and public libraries). It is clear that these various communities differ in their needs (e.g., financing, scale of operation) and roles in society (e.g., population served, desirable breadth of access), as well as in the role that government should play in meeting these needs. As has been pointed out above, the development of a network architecture must be distinguished from its deployment. Also, network research must be distinguished from research in general (research in the various sciences, engineering, and the humanities). Another distinction to emphasize is that between access to a network and access to a service. This distinction is most relevant in considering varying possible levels of access, for example, access to basic telecommunications capabilities and access to information resources available over a network; access to a network does not guarantee access to information services.

NOTES

1. "In barely 12 months, the Internet has gone from a tech novelty to a chic media cliche." See Schrage, Michael L. 1993. "For Time's 'Man of the Year,' Consider the Incredible Internet," *Washington Post*, December 24, p. D10.

2. Computer Science and Technology Board, National Research Council. 1988. *Toward a National Research Network.* National Academy Press, Washington, D.C. (The Computer Science and Technology Board became the Computer Science and Telecommunications Board in 1990.)

3. The Gigabit Testbed program is funded by NSF and ARPA and involves industry contributions and participation as well as participation by universities and government laboratories. See Committee on Physical, Mathematical, and Engineering Sciences, Federal Coordinating Council for Science, Engineering, and Technology, Office of Science and Technology Policy. 1994. *High Performance Computing and Communications: Toward a National Information Infrastructure.* Office of Science and Technology Policy, Washington, D.C.

4. The original four components included NREN, Basic Research and Human Resourc-

es, High-Performance Computing Systems, and Advanced Software Technology and Algorithms.

5. Although commercial networks are often installed in a closed manner for the individual corporation, they share with the Internet the objective of a general communications infrastructure that supports a wide range of higher-level services.

6. The proliferation of business interest can be seen in the rise of special seminars aimed at business use of the Internet and special newsletters, such as "The Internet Letter: On Corporate Users, Internetworking & Information Services" (launched in October 1993), and the *Internet Business Journal* (Canada). Washington, D.C., is the center of Internet-related business activity. "While the Internet has no corporate identity, it does have a center of gravity, and the Netplex [approximately 20 square miles across metropolitan Washington, D.C.] is it. . . . Here reside the dominant for-profit providers of Internet connections and two of the four biggest on-line services, which offer subscribers E-mail, electronic versions of magazines, access to airline reservations and investment services, and the like. The Internet Society is here; so is the computer that houses the master register of affiliated networks' electronic addresses, as well as a raft of Internet services, software houses and hardware makers, and associations. Well over half of Internet traffic between the U.S. and other nations—hundreds of gigabytes each day—passes through the Netplex" (p. 100). See Stewart, Thomas A. 1994. "The Netplex: It's a New Silicon Valley," *Fortune*, March 7, pp. 98-104.

7. For example, revenues from electronic information services for financial management, research, marketing, purchasing, and general business administration were estimated at almost $14 billion in 1993. U.S. Department of Commerce. 1994. *U.S. Industrial Outlook 1994*. Government Printing Office, Washington, D.C.

8. For reference, basic cable subscription revenues in 1993 were on the order of $13 billion; motion picture box office receipts, by comparison, were somewhat over $5 billion, and videocassette sales and rentals totaled between $13 billion and $18 billion. U.S. Department of Commerce, 1994, *U.S. Industrial Outlook 1994*.

9. Roach, Stephen. 1994. "Scoping out the Information Superhighway," *U.S. Investment Research*, Morgan Stanley, New York, February 25.

"None of the experts is suggesting that the information superhighway won't get built, just that early predictions of its quick completion were grossly overstated." Farhi, Paul, and Sandra Sugawara. 1994. "Hurdles Slow Information 'Superhypeway,'" *Washington Post*, April 7, pp. A1 and A15.

10. Roach, 1994, "Scoping out the Information Superhighway."

11. "Not atypically for Hollywood, the most compelling reason many gave for attending the conference was that everyone else was. For the last year, representatives of the communications and media industries have demonstrated an almost compulsive need to trade information highway metaphors at symposiums and conferences with titles such as 'Digital World,' 'Multimedia Expo,' and 'Digital Hollywood.' Although everyone agreed that the highway is coming—and soon—there was little consensus about how it will evolve, how much it will cost or how quickly people will be able to access it through their computers, telephones or television sets. . . . Yet members of the Hollywood crowd seemed unsure of their place in the sparring between phone, cable and computer executives over how digital age entertainment and information will be delivered. 'I feel like an English major in an organic chemistry course,' said Disney's [Michael] Eisner." Lippman, John, and Amy Harmon. 1994. "Gore Presides at L.A. Summit on Info Age," *Los Angeles Times*, January 12, pp. A1 and A19.

12. The ambitious goals and the level of activity initiated by this administration contrast with the more hands-off, market-oriented approach of previous administrations.

13. For example, the requirements for high-definition television (HDTV) to support

emerging NII services are not yet well established. It is known that interlaced displays (today's TVs) were designed for viewing programs at distances of about 8 picture heights (and the aspect ratio is 4 units wide by 3 units high); these dimensions permit little eye scanning by the viewer, which leads to a "disconnected" feeling on his part. On the other hand, HDTV screens have an aspect ratio of 16 units wide by 9 units high, and the viewer is typically only 2 to 3 picture heights from the screen; this permits considerable viewer participation with the activity on the screen. This is consistent with the likely NII applications requiring much closer interactions with the screen. It would be highly desirable if a single HDTV receiver design could be used for HDTV as well as NII applications. To at least some extent, this principle is recognized by the industry partners in the HDTV "grand alliance," which contemplates an "entry-level NII terminal" achievable by upgrading the microprocessors anticipated for HDTV sets. See "National Information Infrastructure and Grand Alliance HDTV System Interoperability," February 22, 1994.

2

The Open Data Network: Achieving the Vision of an Integrated National Information Infrastructure

The committee's vision for a national information infrastructure (NII) is that of an Open Data Network, the ODN. Having a vision of future networking, however, is not at all the same thing as bringing it to fruition. As a contribution to the ongoing debate concerning the objectives and characteristics of the NII, the committee details in this chapter how its ODN architecture can enable the realization of an NII with broad utility and societal benefit, and it discusses the key actions that must be taken to realize these benefits.

An open network is one that is capable of carrying information services of all kinds, from suppliers of all kinds, to customers of all kinds, across network service providers of all kinds, in a seamless accessible fashion. The telephone system is an example of an open network, and it is clear to most people that this kind of system is vastly more useful than a system in which the users are partitioned into closed groups based, for example, on the service provider or the user's employer.

The implications of an open network[1] are that there is a need for a certain minimum level of physical infrastructure with certain capabilities to be provided, for an agreement to be forged for a set of objectives for services and interoperation, for relevant standards to be set, for research on enabling technology to be continued, and for oversight and management. The role government should take here is critical.

The committee's advocacy of an Open Data Network is based in part on its experience with the enormously successful Internet experiment, whose basic structure enables many of the capabilities needed in a truly national information infrastructure. The Internet's defining protocols, TCP and IP, are not proprietary—they are open standards that can be implemented by anyone. Further, these protocols are not targeted to the support of one particular application, but are designed instead to support as broad a range of services as possible. Moreover, the Internet attempts to provide open access to users. In the long term, users and networks connected to such a universal network benefit from its openness. But in data networking today, this vision of open networking is not yet universally accepted. Most of the corporate data networking done currently uses closed networks, and most information and entertainment networks targeted to consumers are closed, by this committee's definition.

THE OPEN DATA NETWORK

Criteria for an Open Data Network

The Open Data Network envisioned by the committee meets a number of criteria:

- *Open to users:* It does not force users into closed groups or deny access to any sectors of society, but permits universal connectivity, as does the telephone system.
- *Open to service providers:* It provides an open and accessible environment for competing commercial or intellectual interests. For example, it does not preclude competitive access for information providers.
- *Open to network providers:* It makes it possible for any network provider to meet the necessary requirements to attach and become a part of the aggregate of interconnected networks.
- *Open to change:* It permits the introduction of new applications and services over time. It is not limited to only one application, such as TV distribution. It also permits the introduction of new transmission, switching, and control technologies as these become available in the future.

Technical, Operational, and Organizational Objectives

The criteria for an Open Data Network imply the following set of very challenging technical and organizational objectives:

- *Technology independence.* The definition of the ODN must not bind the implementors to any particular choice of network technology. The ODN must be defined in terms of the services that it offers, not the way in which these services are realized. This more abstract definition will permit the ODN to survive the development of new technology options, as will certainly happen during its lifetime if it is successful. The ODN should be defined in such a way that it can be realized both over the technology of the telephone and the cable industries, over wire and wireless media, and over local and long-distance network technology.

• *Scalability.* If the ODN is to be universal, it must scale to global proportions. This objective has implications for basic features, such as addressing and switching, and for operational issues such as network management. New or emerging modalities, such as mobile computers and wireless networking today, must be accommodated by the network. If the ODN is to provide attachment to all users, it must physically reach homes as well as businesses. This capability implies an upgrade to the "last mile" of the network, the part that actually enters the home (or business). Further, the number of computers (or more generally the number of networked devices) per person can be expected to increase radically. The network must be able to expand in scale to accommodate all these trends.

• *Decentralized operation.* If the network is composed of many different regions operated by different providers, the control, management, operation, monitoring, measurement, maintenance, and so on must necessarily be very decentralized. This decentralization implies a need for a framework for interaction among the parts, a framework that is robust and that supports cooperation among mutually suspicious providers. Decentralization can be seen as an aspect of large scale, and indeed a large system must be decentralized to some extent. But the implications of highly decentralized operations are important enough to be noted separately, as decentralization affects a number of points in this chapter.

• *Appropriate architecture and supporting standards.* Since parts of the network will be built by different, perhaps competing organizations, there must be carefully crafted interface definitions of the parts of the network, to ensure that the parts actually interwork when they are installed. Since the network must evolve to support new services over time, there is an additional requirement that implementors must engineer to accommodate change and evolution. These features may add to the network costs that may be inconsistent with short-term profitability goals.

• *Security.* Poor security is the enemy of open networking. Without adequate protection from malicious attackers and the ability to screen out annoying services, users will not take the risk of attaching to an open network, but will instead opt, if they network at all, for attachment to a restricted community of users connected for a specific purpose, i.e., a closed user group. This version of closed networking sacrifices a broad range of capabilities in exchange for a more reliable, secure, and available environment.

• *Flexibility in providing network services.* If a network's low-level technology is designed to support only one application, such as broadcast TV or telephony, it will render inefficient or even prohibit the use of new services that have different requirements, although it may support the target application very efficiently. Having a flexible low-level service is key to providing open services, and to ensuring the capacity to evolve. For example, the emerging broadband integrated services digital network (B-ISDN) standards, based on asynchronous transfer mode (ATM), attempt to provide a more general capability for the telephone system than was provided by the current generation of technology, which was designed specifically for transport of voice. The Internet protocol, IP, is another example of a protocol that provides a flexible basis for building many higher-level services, in this case without binding choices to a particular network technology; the importance of this independence is noted above.

- *Accommodation of heterogeneity.* If the ODN is to be universal, it must interwork with a large variety of network and end-node devices. There is a wide range of network infrastructure: local area and wide area technology, wireline and wireless, fast and slow. Perhaps more importantly, there will also be a range of end-node devices, ranging from powerful computers to personal digital assistants to intelligent devices such as thermostats and televisions that do not resemble computers at all. The ODN must interwork with all of them, an objective that requires adaptability in the protocols and interface definitions and has implications for the way information is represented in the network.[2]
- *Facilitation of accounting and cost recovery.* The committee envisions the ODN as a framework in which competitive providers contribute various component services. Thus it must be possible to account for and cover the costs of operating each component. Because the resulting pricing will be determined by market forces, fulfilling the objective of universal access may require subsidy policies, a point discussed in Chapter 5.

Benefits of an Open Data Network

Comparing the success of the open Internet to the limited impact of various closed, proprietary network architectures that have emerged in the past 20 years—systems that eventually either disappeared or had to be adjusted to allow open access—suggests that the wisdom of seeking open networks is irrefutable.[3] Many of the proprietary networks that have played to captive audiences of vendor-specific networks for years are now rapidly losing ground as users demand and achieve the ability to interoperate in a world of heterogeneous equipment, services, and network operating systems. On the other hand, the Internet, and those networks that have "opened up," are enjoying phenomenal growth in membership.

It is important to note that achieving an open network does not preclude the existence of closed networks and user groups. First, there will always be providers (such as current cable TV providers) that choose to develop closed networks for a variety of reasons, such as control of revenues, support of closed sets of users, and mission-critical applications. It is unrealistic to believe that such an approach either can or should be controlled. For this reason, it will be necessary to provide some level of interoperation with proprietary protocols, with new versions of protocols, and with networks that do emerge to deal with special contingencies or with special services. Second, closed user groups will always exist, for reasons of convenience and security. The Open Data Network can be configured to allow closed groups to use its facilities to construct a private network on top of the ODN resources. (See, for example, the discussion below under "Security," which presents approaches, such as the use of security firewalls, to providing a restricted secure environment.)

OPEN DATA NETWORK ARCHITECTURE

To realize the vision of an integrated NII, it is necessary to create an appropriate network architecture, that is, a set of specifications or a framework that will guide the detailed design of the infrastructure. Without such a framework, the pieces of the emerging communications infrastructure may not fit together to meet any larger vision, and may in fact not fit together at all. The architecture the committee proposes is inspired in part by examining the Internet and identifying those of its attributes that have led to its success. However, some important departures from the Internet architecture must be included to enable an evolution to the much larger vision of an open NII.

An Architectural Proposal in Four Layers

Described below is a four-layer architecture for the Open Data Network.[4] The four layers provide a conceptual model for facilitating the discussion of the various categories of services and capabilities comprised by the ODN.[5] The layers are the bearer service, transport, middleware, and the applications.[6]

1. At the lowest level of the ODN architecture is an abstract bit-level transport service that the committee calls the *bearer service* of the ODN. Its essence is that it implements a specified range of qualities of service (QOS) to support the higher-level services envisioned for the ODN. At this level, bits are bits, and nothing more; that is, their role in exchanging information between applications is not visible. However, it should be

stressed that there can be more than one quality of service; the differences among these are manifested in the reliability, timeliness, correctness, and bandwidth of the delivery. Having multiple QOS will permit an application with a particular service requirement to make a suitable selection from among the QOS provided by the bearer service.[7]

The bearer service of the ODN sits on top of the *network technology substrate,* a term used to indicate the range of technologies that realize the raw bit-carrying fabric of the infrastructure. Included in this set are the communication links (copper, microwave, fiber, wireless, and so on) and the communication switches (packet switches, ATM switches, circuit switches, store-and-forward switches, and optical wave-length-division multiplexers, among others). This set also includes the functions of switching, routing, network management and monitoring, and possibly other mechanisms needed to ensure that bits are delivered with the desired QOS.

The Open Data Network must be seen not as a single, monolithic technology, but rather as a set of interconnected technologies, perhaps with very different characteristics, that nonetheless permit interchange of information and services across this set.

2. At the next level, the *transport* layer, are the enhancements that transform the basic bearer service into the range of end-to-end delivery services needed by the applications. Service features typically found at the transport layer include reliable, sequenced delivery, flow control, and end-point connection establishment.[8] In this organization of the levels, the transport layer also includes the conventions for the format of data being transported across the network.[9] The bit streams are differentiated into identifiable traffic types such as voice, video, text, fax, graphics, and images. The common element among these different types of traffic is that they are all digital streams and are therefore capable of being carried on digital networks. Currently much of the work in the commercial sector is aimed at defining these sorts of format standards, mostly driven by workstation and PC applications.

The distinction between the bearer service and the transport layer above it is that the bearer service defines those features that must be implemented inside the network, in the switches and routers, while the transport layer defines services that can be realized either in the network or the end node. For example, bounds on delay must be realized inside the network by controls on queues. Delay, once introduced, cannot be removed at the destination. On the other hand, reliable delivery is normally viewed as a transport-layer feature, since the loss of a packet inside the network can be detected and corrected by cooperating end nodes.

Since transport services can be implemented in the end node if they are not provided inside the network, they do not have to be mandated as a core part of the bearer service. This suggests that they should be separated into a distinct transport layer, since it is valuable, as the committee discusses, to minimize the number of functions defined in the bearer service. While the service enhancements provided by the transport layer are very important, this report does not elaborate further on this layer (as it does below for the other three layers), since these services are a well-understood and mature aspect of networking today.

3. The third layer, *middleware,* is composed of higher-level functions that are used in common among a set of applications. These functions, which form a toolkit for application implementors, permit applications to be constructed more as a set of building blocks than as vertically integrated monoliths. These middleware functions distinguish an information infrastructure from a network providing bit-level transport. Examples of these functions include file system support, privacy protection, authentication and other security functions, tools for coordinating multisite applications, remote computer access services, storage repositories, name servers, network directory services, and directory services of other types. A subset of these functions, such as naming, will best be implemented in a single, uniform manner across all parts of the ODN. There is a need for one or more global naming schemes in the ODN. For example, there may have to be a low-level name space for network end nodes, and a higher-level name space for naming users and services. These name spaces, since they are global, cannot be tied to a particular network technology choice but must be part of the technology-independent layers of the architecture. These somewhat more general service issues will benefit from a broad architectural perspective, which governmental involvement could sustain.

4. The uppermost layer is where the *applications* recognized by typical users reside, for example, electronic mail (e-mail), airline reservation systems, systems for processing credit card authorizations, or interactive education. It is at this level that it is necessary to develop all the user applications that will be run on the ODN. The benefit of the common services and interfaces of the middleware layer is that applications can be constructed in a more modular manner, which should permit additional applications to be composed from these modules. Such modularity should provide the benefit of greater flexibility for the user, and less development cost and risk for the application implementors. The complexity of application software development is a major issue in advancing the ODN, and any approach to reducing the magnitude and risk of application development is an important issue for the NII.

As a wider range of services is provided over the network, it will be important that users see a uniform interface, to reduce the learning necessary to use a new service. A set of common principles is needed for the construction of user interfaces, a framework within which a new network service can present a familiar appearance, just as the Macintosh or the X Window interface attempts to give the user a uniform context for applications within the host. A generation of computer-literate users should be able to explore a new network application as confidently as they use a new television remote control today. Much effort is under way in the commercial sector to identify and develop approaches in this area, and it will be necessary to wait and see if market forces can produce a successful set of options in this case.

A critical feature of the ODN architecture is openness to change. Since the committee sees a continuous evolution in network technology, in end-node function, and, most importantly, in user-visible services, the network standards at all the levels must be evolvable. This requires an overall architectural view that fosters incremental evolution and permits staged migration of users to new paradigms. There must be an agreed-upon expectation about evolution and change that links the makers of standards and the developers of products. This expectation defines the level of effort that will be required to track changing requirements, and it permits the maintainers to allocate the needed resources. The need for responsiveness to change can represent a formidable barrier, since the status quo becomes embedded in user equipment, as well as in the network. *If standards are not devised to permit graceful and incremental upgrade, as well as backwards compatibility, the network is likely either to freeze in some state of evolution or to proceed in a series of potentially disruptive upheavals.*

The Internet is currently planning for a major change, the replacement of its central protocol, IP. This change will require replacing software in every packet switch in the Internet and in the millions of attached hosts. To avoid the disruption that might otherwise occur, the transition will probably be accomplished over several years. In a similar approach the telephone industry has made major improvements to its infrastructure and standards (such as changes to its numbering plan) in a very incremental and coordinated manner. Part of what has been learned in these processes is the wisdom of planning for change.

The committee thus concludes that the definition of an ODN architecture must proceed at many levels. At each of these levels technical decisions will have to be made, and in some cases making the correct decision will be critical to the ODN's success in enabling a true NII. The committee recognizes that if the government attempts to directly set standards and conventions, there is a risk of misguided decision making.[10] There is also concern that if the decisions are left wholly to the marketplace, certain key decisions may not be made in a sufficiently timely and coherent manner. Some overarching decisions, such as the specification of the bearer service discussed in the next section, must of necessity be made early in the development process, since they will shape the deployment of so much of the network.[11] Therefore, both the nature of the architecture's layers and the decision process guiding their implementation are important. In the interests of flexibility, the committee emphasizes the architecture, services, and access interfaces composing an ODN. It describes the characteristics the infrastructure technology should have, leaving to the engineers and providers how those characteristics should be realized.

The Centrality of the Bearer Service

The nature of the bearer service plays a key role in defining the ODN architecture. Its existence as a separate layer—the abstract bit-level network service—provides a critical separation between the actual network technology and the higher-level services that actually serve the user.

One way of visualizing the layer modularity is to see the layer stack as an hourglass, with the bearer service at the narrow waist of the hourglass (Figure 2.1). Above the waist, the glass broadens out to include a range of options for transport, middleware, and applications. Below the waist, the glass broadens out to include the range of network technology substrate options. Imposing this narrow point in the protocol stack isolates the application builder from the range of underlying network facilities, and the technology builder from the range of applications. In the Internet protocols, the IP protocol itself sits at this waist in the hourglass. Above IP are options for transport (TCP, UDP, or other specialized protocols); below are all the technologies over which IP can run.

The benefit of this architecture is that it forces a distinction between the low-level bearer service and the higher-level services and applications. The network provider that implements the basic bearer service is thus not concerned with the standards in use at the higher levels. This separation of the basic bearer service from the higher-level conventions is one of the tools that ensures an open network; it precludes, for example, a network provider from insisting that only a controlled set of higher-level standards be used on the network, a requirement that would inhibit the development and use of new services and might be used as a tool to limit competition.

This partitioning of function is not meant to imply that one entity cannot be a provider of both low- and higher-level services. What is critical is that the open interfaces exist to permit fair and open competition at the various layers. *The committee notes that along with this open competition environment comes the implication that the low-level and high-level services should be unbundled.*

Even if the providers of the bearer service are indifferent to the higher-level standards, those standards should be specified and promulgated. If there are accepted agreements among software developers as to the standards and services at the middleware and application levels, users can benefit greatly by obtaining software that permits them to participate in the services of the network.

Characterizing the Bearer Service

If the ODN is indeed to provide an open and accessible environment for higher-level service providers, then there must be some detailed understanding of the characteristics of this underlying bearer service, so that higher-level service providers can build on that foundation. A precedent can be seen in the telephone system's well-developed service model, which was initially developed to carry a voice conversation. Other applications such as data modems and fax have been developed over this service, precisely because the basic "bearer service," the service at the waist of the hourglass, was well defined.

In the case of data networks, defining the base characteristics of the underlying service is not easy. Applications vary widely in their basic requirements (e.g., real-time video contrasted with e-mail text). Current data network technologies vary widely in their basic performance, from modem links (at 19,200, 14,400, 9,600, or fewer bits per second) to local area networks such as the 10-Mbps Ethernet, to the 100-Mbps FDDI.

A number of conclusions can be drawn about the definition of the bearer service.

- The bearer services are not part of the ODN unless they can be priced separately from the higher-level services, since the goal of an open and accessible network environment implies that (at least in principle) higher-level services can be implemented by providers different from the provider of the bearer service. As long as two different providers are involved in the complete service offering, the price of each part must be distinguished. The definition of the bearer services must not preclude effective pricing for the service. The committee recognizes that such pricing elements have been bundled together and that this history will complicate a shift to the regime advanced here in the interest of a free market for entry at various levels.[12]
- We must resist the temptation to define the bearer service using simplistic measures such as raw bandwidth alone. We must instead look for measures that directly relate to the ability of the facilities to support the higher-level services, measures that specify QOS parameters such as bandwidth, delay, and loss characteristics.

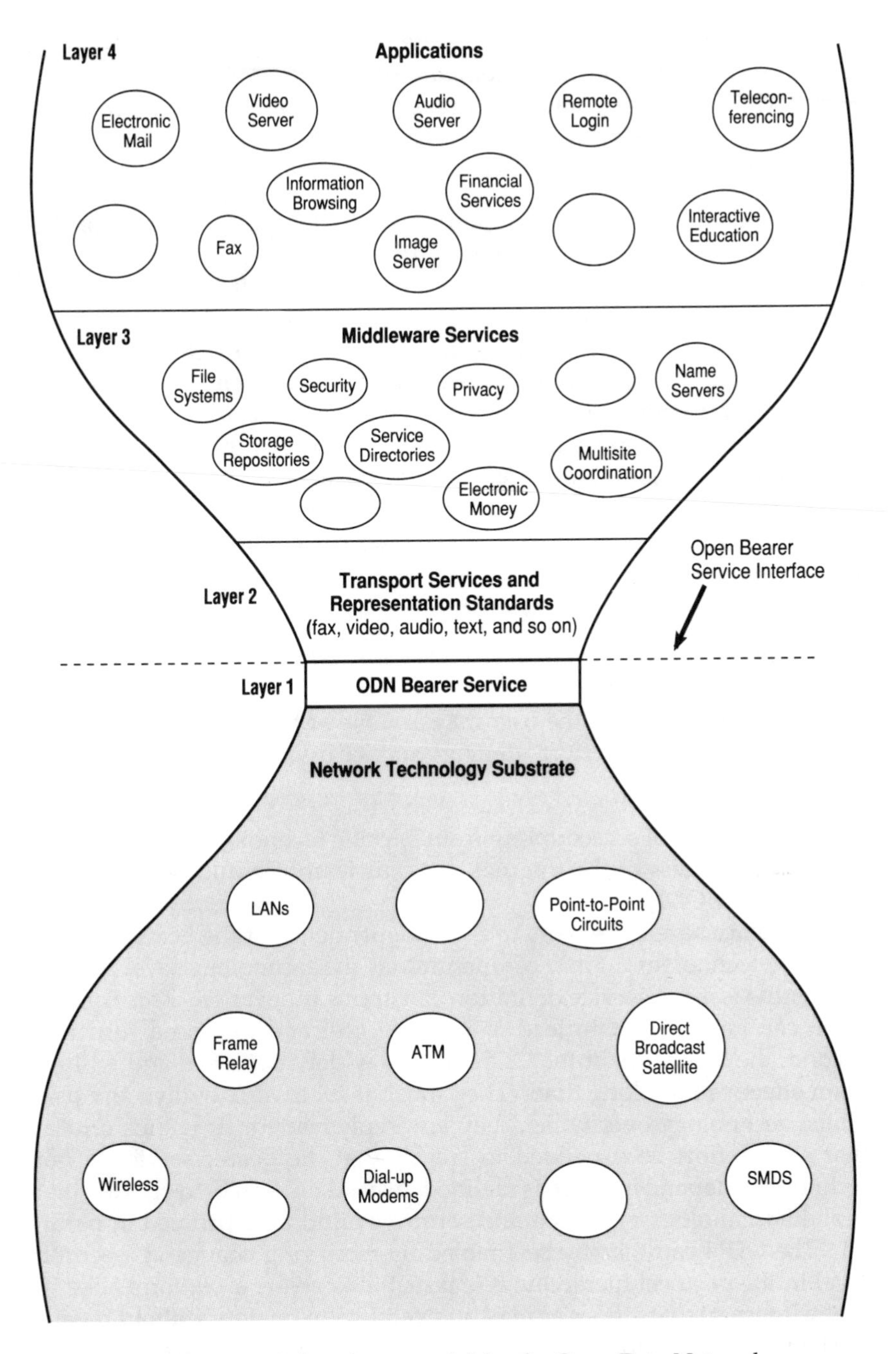

FIGURE 2.1 A four-layer model for the Open Data Network.

• The bearer service must be defined in an abstract way that decouples the service characteristics from any specific technology choice. Network technology will continue to evolve, and the bearer service must provide the stable service interface that permits new technology to be incorporated into the ODN.

This last point deserves further emphasis. It attempts to capture for the ODN one of the key strengths of the Internet, which is that the protocol suite was defined in such a way that it could be realized across a wide range of network technologies with different speeds, capabilities, and characteristics. The analog of the bearer service in the Internet suite is the IP protocol itself; this protocol is deliberately defined in a manner that is as independent from any specific technology as could be accomplished.

This technology independence is accomplished by including in the IP protocol only those features that are critical to defining the service as needed by the higher levels, and leaving all other details for definition by whatever lower-level technology is used to realize IP in a particular situation.[13] IP is thus a very minimal protocol, with essentially two service features:[14]

• The source and destination end-node address carried in the packet, and

• The behavior that the user may assume of the packet delivery service, namely, the "best-effort" delivery service discussed below in "Quality of Service."

The IP protocol's decoupling from specific technologies is one of the keys to the success of the Internet, and this lesson should not be lost in designing the ODN.

There are several benefits to this independence of the bearer service from the technology. First, competition at the technology level will be greater the less the service definition constrains innovation. This competition can be expected to lead to reduced cost and increased function. Second, the standards of the ODN will be widely deployed and will remain effective for a long time. They must, if successful, outlive any particular technology over which they are implemented. It is thus critical that every effort be expended to ensure that the bearer service is not technology dependent, but is defined instead in an abstract way that permits technology evolution and service evolution to proceed in parallel. The ODN can thereby be enabled by managing change at a certain level in the protocol hierarchy; it is possible to create a platform (like IP in the Internet) that allows application vendors to operate without regard to changes in the implementation of transmission lines, switches, and so on—although the platform defines general requirements that designers of transmission lines, switches, and so on should meet. Box 2.1 discusses a specific issue in this context, which is the relationship between the bearer service and a very important emerging technology standard, asynchronous transfer mode.

Middleware: A New Set of Network Services

Above the bearer service and the transport level (layers 1 and 2 of the ODN architecture), but below the application level (layer 4) itself, a new set of service definitions is needed to support the development of next-generation applications. At this level, it is necessary to develop network services for the naming and locating of information objects. Also needed here are tools for buying and selling products (goods and services), including a definition of electronic money, and an architecture for dealing with intellectual property rights. Another necessary service is support for coordinating group activities. Services of this sort have been called "middleware," since they sit above the traditional network transport services but below the applications themselves.

A transition to an environment characterized by an abundance of information of all kinds, forms, and sources implies a need for services to locate, access, retrieve, filter, and organize information. The foundation has already been laid for substantial personalization of information resources (e.g., in the development of personal libraries and the conduct of personalized services). Tools for users are especially important because of the expectation that the provider community is not likely to organize information resources sufficiently, if at all.[15] See Appendix C.

Development of middleware services is not as mature as development of the basic transport services or the specific application services such as those found on the Internet. This area, layer 3, of the architecture is discussed in "Research on the NII" at the end of this chapter.

Electronic money, for example, is a middleware service that illustrates the kinds of issues that must be resolved if the NII infrastructure is to become not just a means for moving bits but also a general environment for on-line services and electronic commerce. One key to the controlled buying and selling of information is an easy means for payment. Today, anyone interested in selling information on the network must either establish a financial relationship with each individual customer or use an information reseller. Information resellers may be an effective mechanism but may also inhibit the freedom to package and structure information in ways that provide product differentiation. A related issue in electronic commerce is that transactions that deliver information, such as directory enquiry, will very often be sold for small sums of money. To bill for each transaction separately would add considerably to the cost of

Box 2.1 The Role of Asynchronous Transfer Mode in the ODN Bearer Service

For the last several years, a proposal has been emerging for a new sort of network technology, which is called asynchronous transfer mode, or ATM. Technically, ATM is a variant of packet switching, with the specific characteristic that all the packets are the same small size, 48 payload bytes. This small size is in contrast to the larger and variable-size packets found in today's packet switched networks (Ethernet, for example, has a maximum packet size of 1,500 payload bytes).

These small packets, which are called "cells" in the ATM standard, provide technical advantages in some contexts. They reduce certain forms of variation in delivery delay, and they make more straightforward the design of highly parallel packet switches, the so-called space division switches.

ATM was originally developed in the context of the telephone industry, as a successor to the current generation of digital telephone standards. In 1988 it was selected by the CCITT (now the ITU Telecommunications Sector) as the agreed-upon approach to achieving a broadband integrated services digital network (B-ISDN).* The perceived utility of ATM subsequently caused it to be standardized in other contexts, such as local area networking.

The main goal of ATM is flexibility. In the current telephone system, a static connection is established and bandwidth reserved from the time a call is placed until it is terminated. Even if a speaker is silent, a flow of (null) bits continues. With ATM (as with packet switching in general), transmission is asynchronous, which means that cells need to be sent only when there is information to transfer. This has the obvious advantage of improved utilization of bandwidth. An equally important point, however, is that this more flexible framework can be used to realize a broad range of services, such as connections of any desired capacity. Today's telephone system is implemented using a fixed hierarchy of link speeds: voice circuits (rated at 64 kbps each) are combined into a circuit at about 1.5 Mbps (called T1), T1 circuits are combined into a circuit at about 45 Mbps (called T3), and so on. In order to sell other speeds (fractional T1 or T3), it is necessary to add additional equipment to the system. With ATM, a circuit of any speed can be configured, by varying the rate at which cells are being sent. This flexibility permits the providers to sell a wider range of transmission services and also provides a basic mechanism for building the internal control and management communications that support the network itself: communications for signaling, for operations and maintenance, and for accounting and billing. Building these facilities with the same technology (multiplexers and switches) that implements the user services has the potential to reduce cost. This situation contrasts with today's telephone network, which has a quite separate data network for signaling.

The other advantage of ATM over traditional telephone circuits is that the physical link from the customer to the network can be used at one time for many simultaneous conversations on which information is flowing in bursts. During the idle period between one burst and the next a conversation consumes no transmission bandwidth and so leaves the transmission line free to carry bursts of traffic belonging to other conversations. By conveying data in small cells, ATM can rapidly interleave bursts of traffic from different conversations and so provide a quality of service that is particularly appropriate for multimedia and computer applications.

As was noted above, ATM may have more in common with packet switching than with the traditional circuit switching of the telephone system. However, in contrast with packet switching as seen in the Internet or on current network technology such as Ethernet local area networks, ATM permits resources (in particular, bandwidth) to be allocated to support specific flows of cells. This feature permits ATM to provide connections with different quality of service (QOS), which refers to distinguishable traffic characteristics such as continuous steady flow, low delay for short bursts, low priority and possibly long delay for bulk-rate traffic, and so on. This explicit control over QOS is in contrast to the single QOS offered by the Internet protocol, the "best-effort" transfer service.

What is most important about ATM is not its technical details, but rather the fact that it has been successful as a common starting point for discussion and cooperation among the previously somewhat disjoint computer, data network, and telecommunications industries. There are now widespread plans for ATM deployment, both in the wide area and local area context. The joint effort put into ATM by these communities represents a force for unification that materially enhances the opportunities for a future integrated Open Data Network.

But while ATM may be a way of realizing the bearer service in the late 1990s, the ATM standards, because they contain many details related to the specifics of ATM technology, do not directly play the role of the bearer service for the ODN described by this committee. They are specific to ATM and cannot describe how ODN traffic should be carried over other technologies such as Ethernet or dial-up telephone lines. As discussions in the text affirm, the committee believes that the ODN bearer service should be defined independently from a specific technology, which will permit both ATM and other sorts of technology to be mingled in the ODN, just as a range of technologies are mingled in the Internet today. To build a network out of heterogeneous network technologies, such as ATM combined with Ethernet, FDDI, and so on, requires a unifying definition of the overall offered service that is independent of the technology details. In today's Internet, this is the objective of the IP protocol; for the ODN this is the function of the bearer service. In the short run, how IP will evolve and how it will interwork with ATM are issues for the Internet and ATM standards bodies. In the longer run, the same issues will be discussed in relating ATM to the broader NII objectives.

continued on next page

Box 2.1—*continued*

*Indeed, the terms "asynchronous transfer mode" (ATM) and "broadband integrated services digital network" (B-ISDN) are often used synonymously. Properly, ATM is a technology approach, and B-ISDN is a standard, but this distinction has been lost, especially since the most active of the current relevant standards bodies calls itself the ATM Forum. Regarding standards setting, "In 1988 there was only a very reduced recommendation with respect to broadband ISDN. It was already agreed then that ATM (Asynchronous Transfer Mode) would be the transfer mode for the future broadband ISDN. Two years later, in 1990, CCITT SGXIII prepared 13 recommendations, using the accelerated procedure. These recommendations defined the basics of ATM and determined most parameters of ATM." See de Prycker, Martin. 1991. *Asynchronous Transfer Mode: Solutions for Broadband ISDN*. Prentice-Hall, Englewood Cliffs, N.J., p. 9.

the transaction. Mechanization by means of electronic money can reduce the cost of billing, making small transactions more affordable and the network more useful.

To address these issues, the middleware layer should provide, as a basic service, the means to effect payment transfer. Were this capability in place, an information seller could deal with any purchaser on a casual basis, trusting in the infrastructure to protect the underlying financial transaction.

The two proposed models for payment are the credit-debit card model and the money model, which differ in functionality and complexity.

A credit-debit card paradigm permits a user to present his identity to a purchaser, with a third party (who plays the role of the authorizing agent in a merchant transaction) assuring the transaction. Although this style of payment is not complex to specify, one of its key aspects is that the identity of the participants is known. This prevents anonymous purchase and allows building a profile of users based on a pattern of purchases. Concerns about privacy are increasing as more and more transactions are carried out on-line, and as requirements grow for users to identify themselves in a robust manner for these transactions.

A more complex scheme for credit-debit transactions is "electronic money," a term connoting a means to transfer value across the network between a buyer and a seller who are mutually anonymous. Such a scheme provides an important degree of privacy and may facilitate wider acceptance of network payment, but it is substantially more complex and more computationally demanding, and it carries the risk of running afoul of existing banking and export regulations.

The problems of network payment are sufficiently complex that the considerable private development efforts under way may not be sufficient. Additional basic research in this area may be needed.

Defining the Higher-level Services

Defining the ODN bearer service and the middleware services is only part of specifying the ODN architecture. There must be broad agreement—derived from the vision of an integrated NII and spanning the range of processes from standards setting to implementation of products and services—on what the core higher-level services will be, and there must be standards that define how these services are to be realized. These agreements and standards are critical to ensuring that users with end-node equipment (computers, "smart" TVs and other entertainment units, and so on) can interwork, even though they are from different vendors and run software from different vendors. The alternative might be that a particular hardware choice becomes coupled to a particular proprietary or immutable set of higher-level services. Again, the objective of the layered ODN architecture is to decouple the low-level technology options (in this case both network and end-node choices) from the higher-level services that can be realized on the technology.

Basic Higher-level Services

Based on its observations of current and emerging uses of the Internet, the committee suggests below a sample list of minimal higher-level services for the future NII. These services should be taken as examples, which may change over time, given the importance of change as a component of openness. The services most relevant to the eventual network may emerge only in the next decade. At present, the committee has identified the following six services as constituting a minimal set:

- *Electronic mail.* The importance of electronic mail, the largest Internet application, is expected to continue. E-mail is mostly textual now but will presumably be multimedia in the near future, as the conversion is already under way.
- *Fax.* Today, fax is transmitted using the existing telephone network and is one of the largest uses of this network. A digital version of fax would be a very low cost service with wide utility. Advanced features of the bearer service, such as real-time delivery, are not needed for fax delivery, as has been illustrated by the emerging transport of fax over the Internet.
- *Remote login: network access to remote computer systems.* The access speed for remote login will increase with time, starting with voice-grade telephone lines or ISDN. Any open network should be required to supply a bearer service adequate to support this minimal level of connectivity, since it is a basic building block for a range of yet higher level services that involve using a remote computer. The standards that define the

remote access should support at least the modes in use today, including emulation of a "dumb" terminal and support for standard window packages such as the X Window system. These services can be supported today over voice-grade lines with compression.

• *Database browsing.* Access to a variety of database services for the purpose of browsing should be permitted. Browsing through digital libraries is one example; another is accessing one's own health or credit records, a service that implies the need for a high level of security assurance. The standards that define the access should provide for at least simple forms of database query operations, which can be packaged by the end-node device in some user-friendly way. Given the exploding success on the Internet of services such as World-Wide Web, Gopher, archie, and WAIS, all of which provide means to browse through and retrieve information (Box 2.2; Table 2.1), it is clear that this area will be of great importance.

• *Digital object storage.* There should be a basic framework for a service that permits users to store digital objects of any kind inside the network and also make them available to others. The term "digital object" denotes an object that is more complex than a file in a file system and that combines contents, type information, and attributes; an example is a video clip. This capability is a first step toward allowing any user of the network to be an information provider, as well as an information consumer.

• *Financial transaction services.* Certain financial transaction services will soon be pervasive. For example, electronic rendering and payment of bills will be a popular service that will enable bills to be directed to an authorized party on behalf of individuals or businesses, and paid electronically. Banks and other financial institutions would handle this function.

These services, which are similar to services now being used or at least explored on current networks, have the characteristic that they are not strongly dependent on high bandwidth or sophisticated QOS support. There are other higher-level services that are more demanding. A partial list follows.

More Demanding Higher-level Services

• *Audio servers.* Audio today is an important component of teleconferencing, multimedia objects, and other more advanced applications. Multicast real-time audio, although now only being experimented with, promises a powerful and compelling service on today's Internet and represents an example of the sort of new service that can be expected on the NII.

Box 2.2 Emerging Information Services on the Internet

The driving application that first defined the Internet was electronic mail. While the Internet also supported file transfer and remote login, e-mail was the service that most people used and valued. It is supported by various directory services such as WHOIS and NetFind.

Recently, we have seen a second generation of applications that are very different from e-mail. Rather than facilitating communication among people, they provide access to information. They offer a means for providers of information to place that information on the network, and they provide a means for users of information to explore the information space and to retrieve desired information elements. These applications—which include archie, Gopher, the World-Wide Web (WWW) and its Mosaic interface, and the Wide Area Information Service (WAIS)—are redefining the future of the Internet and providing a whole new vision of networking.

- Archie is an attempt to make the original file transfer protocol of the Internet, FTP, more useful by organizing the available files and making it possible to search for desired objects. There are, at many places on the Internet, hosts that provide files that can be retrieved by a mechanism known as anonymous FTP—by any person on the Internet without the necessity of having an account or being known to that machine. Although anonymous FTP does permit others to retrieve a file, it is not a very effective means to disseminate information, because there is no way for a potential user to search for the file. There is no way to find all the anonymous FTP sites on the Internet, or to find out if a file is located in any particular spot. Archie attempts to find all the anonymous FTP sites on the Internet, extract from each site a list of all the accessible files, and build a global index to this information.
- Gopher is an attempt to replace the original FTP with a file access protocol that is easier to use—in other words to make it easier to "gopher" the information. In contrast to archie, its focus is less on cataloging and more on easy storage and retrieval. A software package that anyone can obtain and install on a computer, Gopher creates on that machine a "Gopher server," a location in which files can be stored and retrieved. The structure of Gopher makes it substantially easier to list and retrieve files than did previous tools such as anonymous FTP.
- The World-Wide Web (WWW) is a more ambitious project for the representation and cataloging of on-line information. WWW is based on the paradigm of hypertext, in which a document, instead of being organized linearly, is organized as a collection of multimedia objects (typically the size of a few screens of content), each of which has pointers to other relevant objects. The user, by following this web of pointers, can browse in hyperspace without having to be concerned about where the objects are actually stored. In fact, the objects can be stored in any WWW server on the Internet, so that

continued on next page

Box 2.2—*continued*

the user, by following the web pointers, can in fact cross the Internet, perhaps going from country to country as successive objects are explored. The web of pointers, which can be translated automatically by the computer into the desired object itself, make WWW an on-line wonderland for browsing.

The most popular user interface to the WWW is a tool called Mosaic, which displays the web objects for the user to read. The pointer is represented to the user as a highlighted region of the screen; clicking on that region with the mouse causes the object named by that pointer to appear.

Exploration of the top level of the WWW information structure reveals an extraordinary range of information. A recent visit showed, for example, a number of on-line journals in fields as diverse as physics and the classics; collections of pictures, including the Library of Congress Vatican Exhibit (which over a half a million people have visited over the network); and weather maps, guides to restaurants, and personal profiles of researchers. All of these have been contributed by individuals and organizations interested in making their particular knowledge available to others.

This last point is the key to the success of the WWW. The Web is a skeleton, a framework, into which anyone can attach an information object. The Web provides the tools to let anyone become an information producer as well as a consumer. Thus it exemplifies an open marketplace of ideas, in contrast to commercial services that select and control the content of material provided by their systems to users. WWW has been wildly successful. While the size of the Internet has been growing at a rapid, if not alarming, rate, the use of the WWW has grown at an even greater rate (Figure 2.2).

The key to the Web is in the pointers, the linkages that tie one piece of information to another. The pointers from a Web object must be installed by the person creating that Web element, based on that person's expectation of where a user of that object might want to go next. Although there are tools for creating Web objects, the intellectual decisions about what pointers to install belong to the object creator, who thus has great flexibility in how pointers are used. This can lead to very creative and expressive information structures, but also to structures that are rather idiosyncratic.

One consequence of this ability to put pointers into objects is that anyone can build an index to other objects. That is, one can become an information producer not only by adding an object with original content, but also by creating a different way of indexing and organizing the objects already there. The Web thus becomes a testbed for experiments in novel ways to organize and search information.

• WAIS represents a different approach to information searching. In WWW, a human user searches by following Web pointers. In contrast, the WAIS server has used a Connection Machine, a highly parallel supercomputer, to permit very powerful query requests to be made of the server. One of the most interesting requests is to ask the server to develop a word usage profile of a document, and then search all the other documents in the server to find others with a similar profile. If WWW is a means for people to browse in cyberspace, WAIS is a way to let the computer do the browsing.

Although some of these tools are proving very successful, they represent only a first step in the discovery of paradigms for the realization of on-line information. The Web pointers provide a way to navigate in cyberspace but do not provide an effective way to filter a set of objects based on selection criteria. In contrast, WAIS provides a way to filter through all the objects in a server but does not provide an easy way to link objects in different servers.

As Paul Evan Peters of the Coalition for Networked Information observed in briefing the committee, we are in a paleoelectronic information environment, with crude tools, hunters and gatherers, and incipient civilization, but more advanced civilization is coming.* The expectation of far greater quantities and varieties of information combined with far greater ease and diversity in communication implies a need for more effort to develop the necessary related services.

---

*Committee members observe, however, that the Internet experience, with its abundance of information and a rising tide of e-mail traffic, raises the question of whether people will be hunters or prey, and whether people will want to be camouflaged or colorful.

• *Video servers.* Standards are now being defined for the coding and transmission of moving images, such as traditional or high-definition television pictures. These standards will permit services similar to today's cable TV services, as well as advanced offerings such as movies on demand, and the playback of video components of multimedia information services and advertisements. Video will make a range of demands for network services. Delivery of a movie, for example, represents a long-term requirement for bandwidth, while exploring a database of short video fragments represents a very "bursty" load on the network.

• *Interactive video.* While entertainment video can be coded in advance and stored until needed, interactive or real-time video is coded, transmitted, and played back with minimum latency. This form of video is required for teleconferencing, remote monitoring, distributed interactive education, and so on. Coding in real time for interactive video imposes time limits on the coding process that may limit the quality or compression of the signal. A separate standard may therefore be needed for the coding and transmission of interactive video, to support applications such as multimedia teleconferencing.

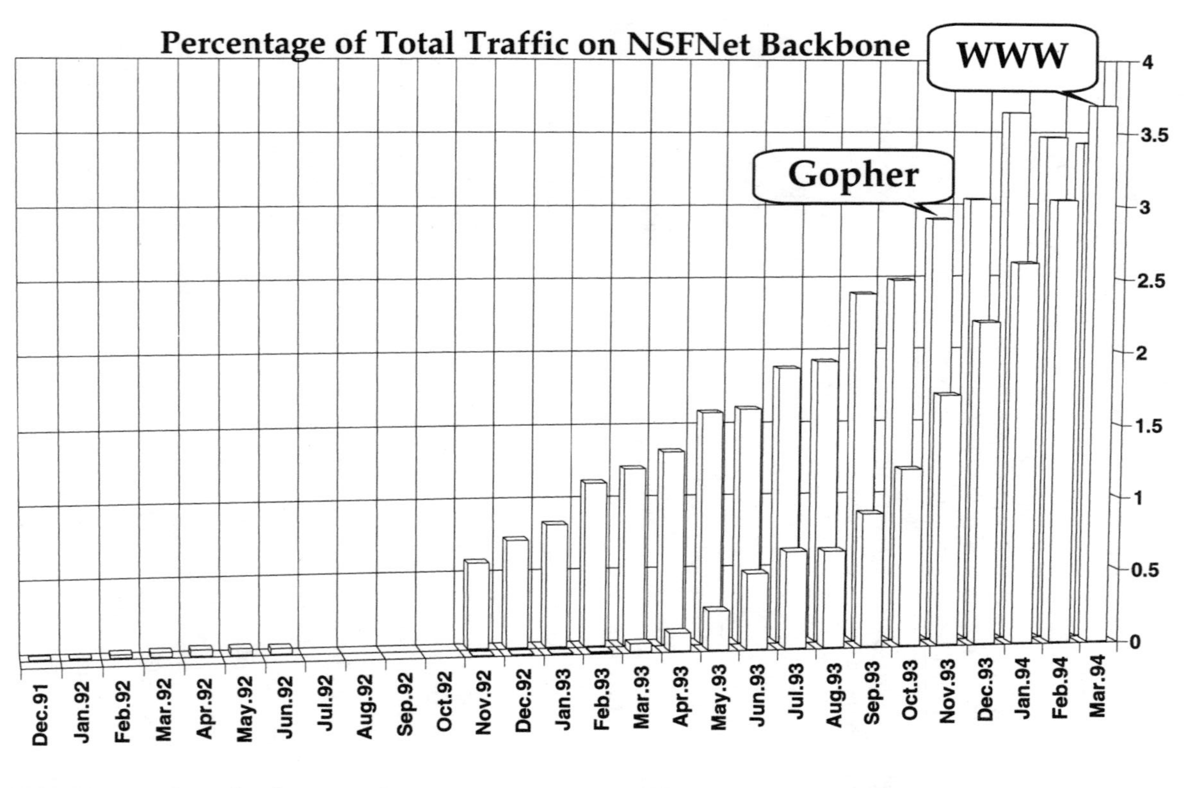

FIGURE 2.2 Growth of Internet browsing services, December 1991 to March 1994. Graph courtesy of the Internet Society, Reston, Va.

TABLE 2.1 Resource Discovery Services

Type	Machines	People	Files	Documents
	X.500	X.500	X.500	X.500
Resource discovery	/etc/hosts NIS DNS	WHOIS finger KIS netfind netdig	archie Prospero Alex Gopher WWW Z39.5	WAIS
Retrieval			FTP NFS AFS Prospero	WAIS Gopher Z39.50
Selection				WAIS Gopher Z39.50

SOURCE: Quarterman, John S., and Carl-Mitchell Smoot. 1994. *The Internet Connection: System Connectivity and Configuration*. Addison-Wesley, Reading, Mass.

QUALITY OF SERVICE: OPTIONS FOR THE ODN BEARER SERVICE

Best-Effort and Reserved Bandwidth Service

An objective of the Internet has been to enable two computers to agree privately to implement some new service, and then implement it by exchanging packets across the network. The only conformance requirements for these packets are low-level matters such as addressing. This flexibility is very important and is captured in the ODN's architectural distinction between the low-level services of the network infrastructure and the higher-level conventions that define how applications are constructed.

The Internet today provides a service that has been called "best effort." When one sends a packet, the network forwards it as best it can, given the other traffic offered at the moment, but makes no specific guarantee as to the rate of delivery, or indeed that the packet will be delivered at all. Many computer applications operate very naturally in a context that does not guarantee bandwidth, and the Internet has demonstrated that best-effort service is attractive in this situation.[16] Just as many operating systems offer a different perceived performance depending on what other processes are running, so also does the best-effort service divide up the available bandwidth among current users. Having an application sometimes run a bit more slowly does not usually cause serious user dissatisfaction. For applications such as remote file access, in fact, best-effort service has proved very effective, as can be gauged by the success of local area networks (LANs) such as Ethernet that provide only this service. Of course, best-effort service is tolerable only within limits, and a network that is totally congested will not be acceptable.

An alternative to best-effort service is one that requires the traffic source to declare its service requirements, so that the network can either reserve and guarantee this service or explicitly refuse the request. When there is excess bandwidth, both best-effort and reserved bandwidth service work well to make users happy. But when there is congestion, the choice is then between slowing all users down somewhat, as is the case with best-effort service, or refusing some users outright in order to fully serve others with reserved bandwidth.[17] For some applications such as real-time audio and video, a reserved service may be preferable.[18] Users will need to be able to express a preference.

This sort of variation in the bearer service is described by the term "quality of service," which is understood in the technical community as covering control of delay, bandwidth, degree of guarantee versus degree of sharing, loss rates, and so on. The Internet today is moving from single, best-effort service to a more complex model with explicit options for QOS, to support new applications such as video and audio. The conclusion in that community is that options for QOS are needed, but not as a replacement for best-effort service, which will remain effective for many applications in the Internet.

The best-effort service and the current cost structure of the Internet are related. The original motivation for packet switching, made available almost 30 years ago, was to permit statistical sharing among users, who were willing to accept the variable delays of best-effort service in exchange for much lower costs. There is no reason to assume that this situation has changed. Some form of best-effort service will probably continue to exist, since it may lead to a more competitive pricing option for many users.[19] Effective bandwidth management will probably continue to be the key to controlling costs in long-lines situations. The NII architecture must provide a range of means to allow bandwidth sharing for cost control. This issue is discussed further in Chapter 5.

Assuring the Service

Whether the offered service is a reserved bandwidth or a best-effort sharing of bandwidth, it will be necessary to offer some assurance that the service actually provided is the one the user was supposed to get.[20]

In the case of an explicit bandwidth commitment, the user can measure the provided bandwidth, but in the case of best-effort service, the user cannot tell if the allocation of bandwidth was actually fair, or if some users somehow received more than they should have.

In the current Internet, allocation of bandwidth is done without robust enforcement. The host software is required to implement an algorithm that regulates the sending rate of the source during periods of overload. But if the end user is clever enough to rewrite his operating system, cheating is possible, since there is no direct enforcement of fair sharing in the switch nodes. Further, only TCP among all the Internet protocols calls for this algorithm; current protocols for voice and video do not adapt to congestion. There is fear in the Internet community that this honor system cannot survive and that some explicit controls will be needed, even in the case of best-effort service, to assure users that the sharing is fair. The Internet Engineering Task Force is working actively in this area. The debate about enforcing fair sharing of best-effort service is likely be a central part of the debate about the bearer service of the NII.

An issue closely related to enforcement is accounting for usage. Usage accounting could provide the basis of billing, a means to assess network allocation after the fact, and a means to assess usage and provide information for planning. A recurring question is whether adding to the current Internet additional facilities for usage accounting would yield sufficient benefits. Feedback to users on actual usage, even in the absence of usage billing, can encourage efficient and prudent use. However, there is a recurring concern that the addition of accounting tools will inevitably lead to usage-based billing, which is in some cases undesirable (see Chapter 5), and a separate concern that the cost of adding accounting tools to the Internet will greatly increase the cost and complexity of the switches and the supporting management environment. The committee notes this matter as requiring further investigation.[21]

NII COMPLIANCE

The previous sections have explored the range of services that could form the basis of the NII, at the low level (the bearer service), the middleware level, and the application level. To define the minimum services that must be provided everywhere in the NII, it is necessary to establish a baseline of defined mandatory functionality. Without some such criteria, any provider could essentially market any network technology or service as being a part of the NII, and users would have no assurance that it would actually be useful. *Defining NII compliance, that is, reaching agreement on the minimum set of NII services, is critical to turning the envisioned NII into an effective national infrastructure.*

However, attempting to define NII compliance in terms of core services that must be available everywhere requires dealing with a basic tension between power and universality. This tension precludes any single definition of NII compliance, so that defining compliance thus becomes rather difficult.

The high-level services discussed above by the committee include some, such as video, that demand substantial bandwidth and sophisticated techniques for traffic management. The committee believes that such services will form the basis of important applications with great utility that will help drive the NII into existence. Yet the necessary bandwidth will not be available or affordable everywhere, even after a significant period of time. To define high-bandwidth services as a mandatory part of NII compliance would thus exclude from the NII infrastructure many components of today's networks, including all the transmission paths using traditional telephone links. Such a restriction is much too limiting. Any realistic plan for the NII must tolerate a range of capabilities, consistent with the different technologies that will be used in different parts of the network.

This reality requires that the basic service of the NII be characterized in an adaptable way that takes into account both power and universality. As a result, the committee's definition of NII compliance has two parts:

- First, an evolving minimum set of basic services, both bearer services and application services, will be required without exception for NII-compliant systems. As discussed above, application services such as electronic mail, fax, remote login, and simple (text-oriented) information browsing do not require advanced infrastructure characteristics and should be available without exception.
- Second, any piece of infrastructure engineered to provide NII services beyond the minimum core will have to be implemented in an NII-compliant manner.[22]

The real consequence of this definition of NII compliance is the maximizing of interoperation, the ability of end nodes attached to the ODN to communicate among themselves effectively, assuring users that any parts of the NII that can support a particular service will implement it in a compatible manner. Users interested in services beyond the minimum set must still verify whether the relevant parts of the infrastructure can actually carry them.

This approach to defining compliance is again motivated somewhat by observation of the Internet. The Internet standards recognize a small core of protocols that must be present in any machine that declares itself Internet compatible. However, most Internet standards, including the essentially ubiquitous TCP, are not actually mandatory, but only recom-

Box 2.3 Testing for Compliance

The term "compliance" carries the perhaps unfortunate implication of a need for formal conformance testing of services and interfaces. In fact, although formal testing seems to work for low-level standards such as link framing protocols, it has not been effective for higher-level standards such as those for electronic mail, remote login, file transfer, and so on. Rigorous testing based on formal protocol modeling has not succeeded in practice in identifying nonconforming implementations.

In fact, the marketplace itself has proved a much more effective test of conformance than has the certification laboratory. In the real world of interworking among heterogeneous implementations, real issues in interoperation are quickly discovered, and vendors who are not responsive to these issues do not succeed.

A specific example of these market forces can be seen in a tradition of one of the first Internet trade shows, called Interop. On the show floor of Interop is installed a real network, and all vendors bringing products to the show are expected to connect to the Interop network and to demonstrate interoperation with their competitors' products. Failures are very obvious and quickly fixed. The resulting definitions of protocol conformance, while perhaps more operational than formal, have served the user community well.

mended. The Internet philosophy is that if a service similar to the one offered by TCP is available on a machine, then TCP should be implemented. This approach has been very effective in practice and seems to represent a middle ground in defining compliance, pushing for interoperation and openness where possible but at the same time accepting the wide range of specific capabilities provided by existing network technology (Box 2.3).

Bandwidth varies significantly over the range of network options, and the question of compliance at various speeds becomes an issue. But bandwidth is not the only dimension in which technology will vary. For example, wireless technology, which is traditionally limited in transmission capacity due to the scarcity of spectrum and issues of cost and transmitter power, may be unable to compete with fiber optics for end-user bandwidth, but fiber systems, which have abundant transmission capacity, lack critical mobility attributes enjoyed by wireless systems.[23] A wireless system would not be deemed noncompliant simply because it failed to support the same bandwidth as a fiber-optic system. Similarly, a fiber-optic system would not be noncompliant because it failed to support mobile users. Each would have to be evaluated within the intrinsic limits of its capabilities. Other issues, such as integrity, privacy, and security, also affect the range of services a particular part of the NII can support.

While it may not be realistic to mandate that support for video and other higher-bandwidth services be a part of the minimum definition of NII compliance, the committee believes that these more demanding services will be key to many useful and important societal objectives in the coming decades. It thus concludes that the bearer services necessary to provide these services should be a long-term objective of the NII.

STANDARDS

Role of Network Standards

To make the vision of an integrated NII a reality and to define NII compliance, it is necessary to specify the technical details of the network. This is the role of standards, the conventions that permit the successful and harmonious implementation of interoperable networks and services. That standards serve to translate a high-level concept into operational terms—and that the process of standards definition is thus key to success in achieving an NII—is well understood in many sectors; indeed, the latter half of the 1980s seemed as much preoccupied with standards as it was with product differentiation. This was true in both the computer and communications industries.

Today, network standards relevant to the NII are being discussed in many different, sometimes competing contexts, such as the following:

- The Internet standards are formulated by the Internet Engineering Task Force (IETF), an open-membership body that currently operates under the auspices of the Internet Society and with support from research agencies of the U.S. federal government.
- The Open Systems Interconnection (OSI) network protocols, which offer an alternative set of protocols somewhat similar to the Internet protocols, have been formulated by the International Organization for Standardization (ISO) internationally, with U.S. contributions coordinated by the American National Standards Institute. A U.S. government version, GOSIP, has been promulgated by the National Institute of Standards and Technology (NIST).[24] ISO is broadening the OSI framework (see Appendix E).
- Standards for local area and metropolitan area networks such as Ethernet, Token Ring, or distributed queue dual bus (DQDB) are formulated by committees under the auspices of the Institute for Electrical and Electronics Engineers.

- Asynchronous transfer mode, an important emerging standard at the lower levels of network service, is being defined by at least two organizations, the ATM Forum and the International Telecommunications Union (ITU) Telecommunications Sector (formerly referred to as the CCITT).
- Standards for the television industry are formulated by a number of organizations, including the Society of Motion Picture and Television Engineers, the Advanced Television Systems Committee, and the ITU.

These sometimes discordant processes are shaped by commercial interests, professional societies, governmental involvement, international negotiation, and technical developments. The recent explosion of commercial involvement in networking has had a major impact on standards definition, as standards have become a vehicle for introducing products rapidly and for gaining competitive advantage.

Factors That Complicate Setting Standards

The committee believes that the critical process of setting standards is currently at risk. Historical approaches to setting standards may not apply in the future, and we lack known alternatives to carry us forward to the NII. The committee discusses below a number of forces that it sees as acting to stress the process.[25]

Network Function Has Moved Outside the Network

As noted above, much of the user-visible functionality of information networks such as the Internet is accomplished through software running on users' end-node equipment, such as a computer. The network itself only implements the basic bearer service, and this causes changes in the standards-setting process. When function moved outside the network, the traditional network standards bodies no longer controlled the process of setting standards for new services based on this functionality. The interests of a much larger group, representing the computer vendors and the applications developers, needed to be heard. This situation is rather different from that in the traditional telephone network, where most of the function was implemented in the interior of the network, and the user equipment, the telephone itself, had characteristics dictated largely by the telephone company. With the advent of computer networks in the 1970s and the strong coupling to computer research, the clear demarcation between users' systems and the network began to blur, and uncontrolled equipment appeared more frequently at user sites, to be attached directly to data networks. Some of this equipment was experimental, and the network could make few assumptions about its proper behavior.

One thing that is clear now is that much of the capability and equipment of connectivity is moving to the "periphery" of the networking infrastructure; for example, there is tremendous penetration of private local area networks on customer premises. This emphasis on the periphery implies that a single entity, or even a single industry, will be incapable of controlling the deployment of the networking infrastructure. Thus the need for an ODN architecture is clear.

It Is Hard to Set Standards Without a Recognized Mandate

The controlled nature of the early telephone system essentially gave the recognized standards bodies a mandate to set the relevant standards. Historically, the telephone network was designed and implemented by a small group that controlled the standards-setting process because it controlled the network. (In most countries outside the United States, the telephone system is an arm of the government.) Similarly, in the early days of the Internet, the standards were set by a body established and funded by the Department of Defense, which (for the DOD) had a mandate to provide data network standards.[26] No such mandate exists in the larger network context of today. For the Internet, for example, the explicit government directive to set standards has been replaced by a process driven by vendor and market pressures, with essentially no top-down control.

A Bottom-up Process Cannot Easily Set Long-term Direction

As can be seen in the Internet community, the absence of a mandate to set and impose standards has led to a bottom-up approach, a process in which the development community experiments with new possibilities that become candidates for standardization after they have been subject to considerable experimentation and use. Standardization is thus akin to ratification of what is the generally accepted practice. The paradigm of translating operational experience into a proposed standard imposes at least one measure of quality on a set of competing proposals. It has proved successful compared to the relatively more controlled and top-down processes occurring in the ISO. But the bottom-up approach to setting standards is not without fault.

Although setting standards by negotiation, compromise, and selection in the marketplace has been largely effective for the Internet, it is important to recall that the Internet had its early success in the context of overall direction and guidance being provided by a small group of highly motivated researchers. This indirect setting of direction seems to have faltered as the Internet community has become larger and more fragmented by commercial interests. Currently, the Internet community seems to make short-range decisions with some success, but long-range decisions, which reflect not only immediate commercial interests but also broader societal goals, may not get an effective hearing. The Internet Architecture Board (IAB) has the charter to develop longer-range architectural recommendations on behalf of the Internet community, but it cannot impose these recommendations on anyone.

A Top-down Approach No Longer Appears Workable

Many people consider the bottom-up approach to be too much like a free-market model in which the final result is due to individual enterprise and competition, that is to say, is not sufficiently managed. The top-down approach appears to be more manageable to many observers, particularly those with extensive experience in managing large-scale networks to meet commercial expectations for performance. However, the classical top-down approach has not succeeded in the current environment, whereas the bottom-up Internet process, which has directly embraced diverse approaches and objectives in its bottom-up process, is a phenomenal success that must be applauded and respected. Although there may be merit in considering how to integrate the top-down and bottom-up approaches, there is little experience to suggest that either approach alone will work easily in the larger context of the future NII.

Commercial Forces May Distort the Standards-Setting Process

A vendor, especially one with a large market share, can attempt to set a unilateral standard by implementing it and shipping it in a product. Some of the most widespread "standards" of the Internet are not actually formal standards of the community, but rather designs that have been distributed by one vendor and accepted as a necessity by the competition. This approach can lead to a very effective product if the vendor has good judgment; it may open the market to innovation and diversity at a higher level based on the standard. However, the objectives of the vendor may not match the larger objectives of the community. The resulting standard may be short-sighted, it may be structured to inhibit competition and to close the market, it may simply be proprietary, and it may inhibit evolution.

Setting Standards for the NII—
Planning for Change Is Difficult But Necessary

Managing the process by which the NII network environment evolves is one of the most critical issues to be addressed. Planning for change requires an overall architecture and constant attention to ensure that any standard, at any level of the architecture, is designed to permit incremental evolution, backward compatibility, and modular replacement to the extent possible. Unfortunately, as the committee has noted above, the current standards-setting processes seem least effective in setting a long-term direction or guiding the development of standards according to an overall vision of the future. Thus there is some need to find a middle ground whereby an overall vision of the NII can inform standards selection and also allow for competing interests and approaches to be evaluated in an open process. *The critical question is not what the exact vision is, but how it will be promulgated and integrated into the various ongoing standards activities.*

ISSUES OF SCALE IN THE NII

The committee views the NII as being universal in scope, reaching not just to businesses and universities, but also eventually to most homes, as does the telephone system. This objective raises many issues related to growth and scaling.

A major research focus of the last 10 years has been scaling network bandwidth to higher speeds (gigabits and beyond). Scaling in the number of nodes has perhaps received less attention. But the issues of scale in the number of nodes are perhaps more challenging that those of speed. As the network gets larger, issues of addressing, routing, management and fault isolation, congestion, and heterogeneity become more relevant. These issues are further complicated by the likely decentralized management structure of the NII, in which the parts of the network will be installed and operated by different organizations.

We see in the Internet today that some of the protocols and methods are reaching their design limits and need to be rethought if we are to build a network of universal scale. A major effort is now being made to deal with serious limitations in the Internet's current addressing scheme, for example. Close attention should be paid to the resolution of these problems in the Internet to derive insights for the far larger NII.

Addressing and Naming

The Open Data Network envisaged by the committee will surely grow to encompass an enormous number of users and will be capable of interconnecting every school, library, business, and individual in the United States, extending beyond that to international scale. The ability to communicate among such a huge set requires the ability to name the desired communicant. The Internet currently provides a 32-bit address space, within which it is theoretically possible to address approximately 4 billion hosts. However, as a result of the structured way in which this address space has been allocated, the number of addresses may be restricted to far fewer than 4 billion, a limitation that the Internet standards community is working to rectify for both the short and a longer term. Similarly, the international telephone numbering plan is facing the result of the tremendous growth in demand for services and lines caused by the advent of fax, cellular phones, and soon personal communicators and other highly mobile services. In the case of telephone numbering, the demands of international commerce have further aggravated the problem by calling for universal information and 800-number services. The Internet and the telephone naming systems' simultaneous arrival at a crisis suggests that perhaps a common solution can be developed.[27]

The current address spaces of the Internet and the telephone network are a low-level framework suited for naming network and telephone locations and delivering data and voice. This framework is not used for naming users of the network or service providers or information objects, all of which must be named as well. Developing suitable name spaces for these entities involves issues of scale, longevity, and mobility. User names, for example, should be location independent (more like unique identifiers than locators), which in turn implies a substantial location service that translates names into current locations.

The current services in the Internet lack a suitable naming system for users. One of the most user-friendly actions the NII community could take would be to rectify this situation. Past attempts have failed for a number of reasons, mostly nontechnical, and this lack of success perhaps deters the community from trying again. In this and other situations, a leadership with a mandate could stimulate useful long-term action to achieve needed results.

Clearly the naming problem will grow substantially as more addressable (and nameable) devices proliferate; it has been suggested that numerous electronic devices (such as thermostats and stoves as well as devices explicitly intended for computing and communications) in homes and workplaces could be connected to networks. Some devices may be mobile, some stationary. Thus it is important to remember that names can be used to identify people, devices, locations, and groups.

The migration to a new address space will be a major upheaval that will affect users, network providers, and vendors. Unifying the various network communities, the Internet, the cable industry, and the telecommunications industry is an additional complex undertaking that will not happen unless there is a clear and explicit advantage. An effort to define a single overarching architecture is the only context in which this integration can be motivated. It will take careful consideration to plan and implement a scheme that properly resolves such major concerns as an appropriate addressing scheme, interim management actions, and migration plans. This implies that overarching architectural decisions for the NII, such as addressing, must be made in a context with an appropriate long-term vision and architectural overview.

Wide-ranging discussion is needed on the requirements for next-generation integrated addressing, with the goal of determining what the scope of this coming address space should be. It is critical that the requirements be appropriately specified. The government and the existing private-sector standards bodies should cooperate to ensure that the resulting decision meets the needs of the future NII. *Sustaining a discussion among the players about core issues such as addressing, and then creating a context in which the resulting decisions are actually implemented, is an example of an effort in which government action could have significant payoff.*

Mobility as the Computing Paradigm of the Future

We are in the midst of a subtle, but powerful, change in the way individuals access data and information processing systems. It is now the case that end users are often "on the road" or simply away from their "home" or "base" computing environment when they need access to that environment. That is, the computing infrastructure in this country is rapidly becoming mobile. One-third of all PCs sold today are mobile, a fraction that is rapidly approaching one-half. Moreover, more than 30 million people now have computers at home as well as in the office, and they often conduct business on their home machines. Thus a new paradigm is emerging in which location-independent access to personal and shared data, resources, and services will be the goal of our information processing infrastructure. Although relatively small systems to support mobility have been built, the complications will increase dramatically with the enormous number of nodes envisioned for the NII.

Mobility raises a number of issues that have to do with the software architecture. One problem is the uneven capability of mobile systems. For example, the computers in use will include today's high-performance laptop computers, as well as intelligent terminals with a quality user interface. The available communications infrastructure will vary greatly over time and from place to place, spanning the range from gigabit bandwidth and microsecond delays to nearly no bandwidth and delays measured in hours or days.

It is unreasonable to ask that the application handle such variability by itself. Rather, the architecture must provide software that can support mobility. From the user's perspective, the network should present to the application the appearance of a coherent global environment, which does not depend on the quality of the available communications, and which assures the availability of needed resources. Examples of some of the elements of the underlying system support functions are predictive caching, dynamically partitionable application architectures, and an appropriate naming and addressing scheme.

Although issues of mobility arise with or without wireless connectivity, it is important to recognize that wireless versus wireline connectivity has other important repercussions. Issues related to privacy, security, authentication, bandwidth, addressability, and locating make it clear that the architecture developed for an NII should not be limited to that of wireline connectivity. Full development of a proper model of mobility has yet to be conducted and is certain to be a critically important area of investigation for the emerging NII.

Management Systems

In a network and service aggregate as complex as that envisaged for the NII, it is critical that network management be attended to properly. The NII will require continuous support to keep it up and running and to ensure that users have a high level of available service. Since the NII is likely to be a conglomeration of many interconnected networks, the network management function must be able to interoperate in a highly heterogeneous environment that is in a continual state of change, growth, and improvement. It is likely that this management function will be distributed across the component networks, a difficult reality with which we must be prepared to deal. This is not a new problem; industry today is paying considerable attention to management systems.

Measurement and Monitoring

Measurement of a network's behavior is important to understanding its functioning. As the network grows very large and becomes very distributed, the process of data collection, the bandwidth needed by monitoring tools, the requirements for storing and processing of the collected data, and so on are all affected. Techniques to detect faults and to diagnose and predict performance are essential, as is the ability to do automated or semiautomated measurement, reporting, and even repair in some cases.

However difficult detection, diagnosis, and repair may be in the context of a given network, the difficulty is compounded in an NII environment. One network's occasional loss of a packet, much less the source of the difficulty, may be difficult to pinpoint. Dealing with intermittent conditions that occur in an environment not administered by a single entity is among the more difficult of the technical challenges to be faced. Another difficult problem is characterizing and controlling the performance of a complex system consisting of numerous independently owned and operated components.

In order to monitor the behavior of traffic and users for purposes of billing and activity summaries, it is important that measurement hooks be judiciously placed in the network components (switches, routers, line drivers, and others). Another critical function that can take advantage of measurement capabilities is network control. Because networks are occasionally subject to disastrous failures that bring down major functions in society and industry (recall the past effects of power failures on Wall Street or of telephone outages on airline reservation systems, air traffic control, or 911 services), it is imperative that controls be placed on network traffic to avoid such catastrophes. The kinds of control the committee refers to—admission, flow, routing, error, and congestion controls, among others—will be designed to react to traffic overloads, network dynamics, network hardware, software failures, and so on. All of these control issues are the subject of current research, and many approaches are being tested in the various networking testbeds.

In addition to its use in network management and control, measurement is essential to evaluating a network's performance. The various measures of performance include response time, blocking, errors, throughput, jitter, sequencing, and others. How measures of these functions affect users' perceptions of satisfactory service is a matter of great interest to the providers of network services and infrastructure. Data on measured actual performance form the basis of models that can be constructed to predict the operation of the network over a wide range of system parameter values.

SECURITY AND THE OPEN DATA NETWORK

Securing the Network, the Host, and Information

There are certainly advantages to having a ubiquitous open data highway. The telephone system, for example, is ubiquitous in the sense that attaching to the telephone network makes it possible to reach and be reached by any telephone. Society has come to depend on universal telephone connectivity, and computer network connectivity can be equally useful. However, the very real threats faced when computers are reachable on the network—threats such as theft of data, theft of service, corruption of data, and viruses—must be anticipated. This has led many companies to hide their corporate networks behind special-purpose gateways, such as mail gateways, that allow only a limited set of applications to cross into or out of the network. Another solution has been the closed user group; some companies have used networks to interconnect their sites, but they do not want connectivity between any of their sites and anything else. Many public provider networks provide such functionality.

The openness of the telephone network is occasionally a detriment, as with prank calls, telephone harassment, and unwanted telemarketing. Most people are sufficiently resistant to these annoyances that they tolerate them in exchange for the utility of the network. But computers seem more vulnerable than people, and the effects of compromised computer software can be catastrophic. Computer security is not a solved problem, and virtually all systems are vulnerable to break-ins and viruses.[28] A virus can render a computer and all data on it unusable.

Developing a Security Architecture

If the NII is to flourish, we must provide solutions so that any end node attached to the network can mitigate its risk to an acceptable level. Further, the infrastructure must be developed to achieve sufficiently high reliability, and the architecture must protect against system-wide failures. Infrastructure security and reliability must also address the vulnerabilities inherent in wireless technologies.[29]

The prospect of new services and applications emerging on the NII increases the urgency with which security must be addressed. Electronic commerce will increase the threat of fraud. Accounting for network usage may lead to theft of service. These threats may require better tools for user authentication, which may in turn lead to increased concern about the privacy of on-line activities. Mobility may lead to new location-independent network names, which may raise similar concerns. Gateways used to restrict network access to known services such as electronic mail will prevent the introduction of new services and applications. All of these matters will arise as the network evolves in the next few years, and all must be addressed now.

The committee sees security as another area in which active governmental involvement can materially advance the state of the NII. As part of its overall support for the NII, the government should foster the development of a security architecture.[30] This architecture should provide for mechanisms that protect against classic security threats (to confidentiality, integrity, and availability of data and systems) as well as violations of intellectual property rights and personal privacy. This security architecture must include technical facilities, recommended operational procedures, and means for recourse within the legal system.

The most well developed security architecture is that used by the

Department of Defense for the protection of classified information. This model does not seem adequate for the full range of problems to which the NII will be subject, and research on a broader spectrum of security problems is needed. Currently, the federal government, particularly the Advanced Research Projects Agency (ARPA), is funding research in robust, available networking. The committee supports this approach and urges other agencies such as the National Science Foundation (NSF) to have an explicit program in this area.[31] Three elements of security issues that are not well characterized at the present time, and that seem critical to achieving success in the NII, are as follows:

- First, effective protection of data and systems will require the use of secure "walls" to separate network functions and service offerings that are expected to be accessible from those that are not. The network must allow information providers to determine the degree of access that will be permitted to their works. The architecture must allow these walls to be constructed so that controlled access through the walls can be implemented.
- Second, technology will have to aid in protecting data integrity. It is critical for information creators to know that what they produce is what network users get, and for users to be assured that what they are getting is what they think it is. There must be protection against the dissemination of a work altered without authorization. A technological means is needed for "certifying" the authenticity of the data, so that users are able to choose sources of information with a reasonable degree of confidence.
- Third, the network itself must provide the reliability (availability) that is critical to the delivery of any higher-level service. Today, users view the Internet as reasonably stable, but it is known that it can be seriously disrupted by abuse from an end node, either by gross flooding of the network with traffic or by the injecting of false control packets that disrupt the internal state of the network, for example, the routing tables.

There is evidence in the network communities of increasing concern about security, concern heightened by reports of incidents and abuses.[32] For the last several years, all proposals for Internet standards have been required to include discussion of impacts on security. Although this requirement does force the community to notice in passing the issue of security, the committee urges an even stronger emphasis on and expectation for security requirements in key network protocols.

Security Objectives and Current Approaches for Reaching Them

It is often assumed that security problems can be solved by the creation and deployment of new security technology. While some technological means exist that currently are not fully exploited, the problem is not just technical. Ensuring security requires attention to operating procedures, user attitudes and values, policy and legislative context, and a range of other issues.[33] The following sections discuss existing technology in the larger context of particular security objectives.

Computer System Protection

It is possible to imagine that individual computers and operating systems could be made secure, but current practice limits the ability of today's products to offer robust protection. Any host attached to a network such as the Internet must assume that it is vulnerable to attack by an attacker that can penetrate the computer. Although attacks are not a common problem on the Internet, their occurrence must be anticipated.

Apart from the assurance of better end-node security, the current approach to ensuring computer system protection is to impose a computer in the network that restricts the range of access. One method is to use a mail relay, which certainly restricts the range of access; unfortunately it does this so well that it prevents applications other than mail from succeeding. The other, use of a router that forwards packets only between a certain set of hosts, can permit a wider range of services but must occur in combination with some flexible approach to specifying the hosts that are permitted to communicate. Although such router technology exists, there is no generally accepted architecture for administering the control that it can provide.

Protection of Information in the Host

Given that system penetration cannot be totally prevented, it is necessary to have some means to prevent damage to information in the host as a result of penetration. To protect from loss, the best method is the old-fashioned one of backup to detached media. This approach protects from both attack and physical failure and should be a standard procedure for almost any computer user.

Protection against disclosure and corruption is more subtle. Disclosure can be protected against in several ways. One, the model used in the military and intelligence communities, is to implement, as part of the system, an inner core of protected mechanism (a kernel that monitors access to user information) that is intended to survive during a system penetration and that will maintain control of access to the information. Another model, so-called discretionary control, typically involves user authentication and access control lists to identify the users permitted to access information objects. This sort of control is widely used in systems today, but many in the security community consider the typical access control list mechanism to be less robust than the security kernel. A final means of protection, which does not depend on system software, is encryption of the information when it is not being used. This method is effective with two limits. First, the information must be transformed to its unprotected form to be used, and this opens a period of vulnerability. Second, encryption depends on having and protecting the encryption key, which is not easy. If the key is stored in the computer, then it is no more protected than the original file was. If it is stored outside the computer, it may be forgotten or compromised by physical means, such as by being observed on the paper on which it is written. Without care, an encryption system becomes no more secure than a password system.

The committee recognizes that the various uses of encryption are current matters of government policy discussion.[34] Such matters as export control, and whether keys must be made available to the government, will have an impact on the ways in which encryption is employed.

Protection of Information in the Network

As commercial traffic grows on the Internet, so also will the temptation to attack that traffic. The only effective means to protect information in the network is encryption. There is no direct equivalent of the military security kernel that can protect data even in the event of a penetration. Encryption makes data unreadable to attackers without the encryption key: thus it protects from all the gross forms of attack and provides a clean separation between those who should have access and those who should not. Again, the problem is protecting the key used to transform the information, but the difficulty is greater in the network because the key must be shared between the sender and the receiver.

There are two solutions to the problem of protecting the key: a systematic way of managing shared keys, and "public-key" systems. Among a small group of cooperating users, it is reasonable to imagine a reliable way of sharing keys off-line, such as mailing them printed on paper. But this approach is neither fast nor easily scaled to global communication.

Public-key systems, sometimes called asymmetric systems, attempt to reduce the problem of key distribution by implementing a scheme in which a separate key is used to encrypt and decrypt. This separation means that the encryption key can be widely publicized (e.g., listed in name servers), while the decryption key is kept secret. Thus, anyone can encrypt a message, but only its intended receiver (with the secret decryption key) can decrypt it. This technique seems a very powerful way to address the problem of keys.[35] It does not entirely eliminate the concern with keys, however. For example, the decryption key can still be stolen, or the sender can be tricked into using the wrong encryption key, so that the wrong recipient can read the message.

Authenticating Users

Most security controls depend on some way to verify users' identities: the user authentication process. The need to identify and authenticate users exists at many levels of the security architecture. At the network level, it is necessary to identify users for purposes of accounting and billing. At the system level, it is necessary to authenticate users so that they can be given the proper authorization. The traditional access control list implies that the system can know which user is making a request, so that the proper entry on the list can be evaluated. Network applications such as network mail also would benefit from a system that would allow a mail recipient to verify the sender's identity. Systems with this capability, such as the Internet's Privacy Enhanced Mail, are just now being deployed.

The most common method for authenticating users is to demand a password from the user as a proof of identity. Although it is well understood, this scheme has a number of weaknesses, both in the human factors (people forget the password, or pick a password easily guessed, or write it down in a visible place) and technically (some network technologies make wiretapping very easy, which facilitates casual theft of passwords, as a recent rash of attacks on the Internet has illustrated).[36] Today, systems are being deployed that use cryptography rather than simple passwords to provide a more robust facility. These systems are beginning to be used on the network, but they are not yet widely integrated into commercial systems.

Control of Authorized Users

While encryption can be used to keep data from the hands of unauthorized users, encryption in its basic form offers no limits on what an authorized user can do. Once the user has the key to reveal the contents of a file, there seem to be no limits as to what that user can do to it. In particular, the user can copy the information, change it, and give it to other users. There is no identified technical means to prevent this copying, without unreasonably limiting the primary use of the information, although technology is under development in this area.

However, encryption can be used to achieve some specific capabilities. Most importantly, cryptography can be used to provide a certificate, a digital signature, which attests that a file's contents have not been corrupted. Such a certificate cannot be forged, and so it provides assurance from the creator that a file is intact.

Taking a Comprehensive Approach to Ensuring Security

Technology provides some effective tools for use in building a secure and effective system, but it is not the whole answer. Technology must be made part of a total approach that includes a set of operating assumptions and controls that allow the users to make reasonable decisions about the operation of a system. If, for example, all users passing along a piece of information were required to attach an additional certificate, asserting the integrity and authenticity of the contents, then presumably there would be an incorruptible trail from any receiver back to a source that claimed to be the original creator. Such a trail would not prevent data theft and corruption, but would provide evidence of the event. However, this control would be effective only if users understood that they should not be party to receiving data without a certificate. Such a significant departure from current practice would doubtless generate great resistance. But implementation of these practices by major vendors of software for information handling might help the scheme to succeed.

The committee concludes that progress in the area of trustworthy and controlled dissemination of information does not depend primarily on technology but rather on the development of an overall model, or architecture, for control, as well as education and public attitudes that promote responsible, ethical use of information, and associated regulation and policy. Although this model can make effective use of technology components, it will derive its strength from its acceptance by the community. At the same time, there is still a strong need for research and development in the security area, both to develop new technical concepts in key areas and to explore alternative approaches to architecture and operations.

FINDING AND BALANCING OPPORTUNITIES TO BUILD TOWARD CONVERGENCE

As the committee has pointed out in Chapter 1 and above, a diverse but interrelated set of expectations and requirements is driving current efforts aimed at developing a U.S. national information infrastructure. One vision of the NII, developed around the notion that it will provide citizens access to the information they need over a ubiquitous national network, derives from the observed success of the Internet and includes open access and a range of services. Another, the combined entertainment, telephone, and cable TV (ETC) vision, emphasizes access to a large number of television channels providing video entertainment and other related services for which there is perceived to be a large market that promises considerable commercial success. Harmonizing these visions conceptually and technically will require finding specific points at which convergence can be encouraged. Competing needs and interests will have to be balanced as we take opportunities at hand now to begin to construct an open NII that will serve the nation well into the future.

Development of Standards for Television—An Example

A particularly important opportunity exists in the development of new standards for television, including high-definition television (HDTV). The move to HDTV represents a move from the old analog standard of NTSC video to a digital standard for coding and transmission. A key standards issue is how to relate methods for HDTV coding and transmission to the broader matters of direct importance to the NII.[37] There is no question that video will be an important component in the emerging NII, and it will be packaged and transmitted along with data and other digital information according to various standards. If broadcast channels (including those that are now envisioned to be made available only for HDTV) were to support a wide range of applications including, but extending well beyond, HDTV, then they would become general-purpose digital distribution channels, which could couple at their destinations into networks such as LANs that are designed to further transport the information.

Although the current HDTV architecture is layered, it is not really modular; that is, the interfaces between layers are not defined with the goal in mind of allowing different layers to be replaced with other layers as technology and distribution methods evolve and/or dictate. The main goals of the current HDTV design were limited to the primary issues of concern—to support television with higher resolution and to efficiently use broadcast television channels with high-resolution signals. It is important to expand these goals in light of the emerging NII and its applications.

The same considerations apply in the nearer term to the evolving cable TV infrastructure. At present, digital TV channels on next-generation cable TV systems are assumed to be tailored to the delivery of television using the MPEG-II standard. Unless the need is recognized for a general service capable of delivering digital information of many forms, the details of the cable infrastructure may become so targeted to MPEG-II that no other information can be added to the system without extensive modification.

The economic forces that shape the reconstruction of the infrastructure are very compelling. Generality in a system almost always adds to cost. The future television standards represent an excellent example of the tug-of-war between generality and specificity. If the future standards for video dissemination are specifically engineered for the delivery of TV only, cost optimizations may result that are in total very substantial, due to the large number of expected end nodes and due to economies of specialization; such economies support an argument for multiple outlets or interfaces specialized to individual applications. But such a monetary cost reduction might come at the opportunity cost of preventing any of the broader goals of the NII (notably integrated service access and delivery) from being implemented over this infrastructure. This, surely, would be a greater cost. Nevertheless, the television, cable, and related entertainment industries cannot be expected to take on themselves the objective of engineering into their systems extra cost to meet goals that lie beyond their objectives.[38] It thus seems likely that market forces alone will not produce an access infrastructure that can accommodate diverse uses over time.

Reengineering of the Nation's Access Circuits

Events as they are currently developing suggest that cable and telecommunications industries vary widely in their recognition of the vision of an integrated NII as articulated by the committee. Although some isolated service offerings are very exciting (such as the proposal from Continental Cablevision and PSI to offer Internet access service),[39] there is no consensus either as to the importance or the viability of a broadly useful NII.

A key ingredient in this lack of consensus is that there is no widely agreed upon method as to how, or if, cable infrastructure will attach to the desktop computer in the home, to the data networks that exist (or are being planned), to business environments, and so on. This is an urgent matter to resolve in the context of the NII. It is particularly important in the context of standards and interfaces for consumer equipment, which is very difficult to upgrade or replace once it is widely deployed—as the example of NTSC, the analog coding standard for broadcast television, makes clear. The practical impossibility of replacing all the televisions in the country is causing the NTSC standard to have an impact and consequences far beyond its technical merits.

A major example of commercial interests moving ahead very quickly without a context in which to address broader public interests is now in the making, namely, the currently planned reengineering of the nation's access circuits. Articulation of public interests, preferably with direct government involvement, could help to shape these developments. The committee recognizes that, as this report was being written, the Technology Policy Working Group of the Information Infrastructure Task Force was contemplating relevant activities.

Cost and Function in Access Circuits

One of the key issues in the argument over the reengineering of access circuits is the potential of added cost to provide reasonable bandwidth going away from the home, the so-called "back" or reverse channel. There are several apparent trade-offs in this area.

One of the committee's key objectives is that the overall structure of the NII not act to preclude anyone from playing a particular role in offering a high-level service. For example, it should be possible for any user to become a provider of information on the network. This objective is an issue because in many cases producers and consumers have asymmetric needs for bandwidth; the high-bandwidth path runs from the producer to the consumer, with only a low rate of data transmission in the reverse direction. Today some networks provide symmetric data paths, whereas in other networks the paths to and from the user are very different. For example, the telephone network provides symmetric bandwidth to and from the user, while the cable networks provide vast bandwidth to the user, but little or no return communication.

Such asymmetry is a potential limitation to the development of an open NII, because it means that becoming a producer of information (whose communication into the network requires large bandwidth) is possible only by entering into a special agreement with the network provider. Normal users can only be consumers of information, and the network provider controls access to what the producers provide.

The objective of allowing for equal two-way communication must be balanced with the technical and economic problem that symmetric channels may not be cost-effective for many situations; a related issue is to assess the costs and benefits of back channels of varying sizes.[40] Engineering high-bandwidth reverse channels into an existing cable TV system would add substantially to the system costs. These costs are hard to justify in immediate practical terms, since many access paradigms are asymmetric at the application level. For example, a great deal of asymmetry is built into the architecture of the popular client-server model, and access to certain archival library databases implies asymmetry as well. On the other hand, peer-to-peer applications can often exploit symmetrical channel access.

It is important, however, to recognize the value of introducing into asymmetric networks the potential for symmetry, even if that capability is not universally deployed. That is, the overall architecture should provide for symmetry although individual applications need not. Indeed, the issue should not be choosing asymmetry as opposed to symmetry, but rather determining how to provide flexibility so that customers can purchase the appropriate degree of asymmetry.

Options for Incorporating the ODN Bearer Service

There are a number of different options for incorporating the committee's concept of the open bearer service into an access network intended primarily for some particular application, such as delivery of cable video. The five different patterns of integration discussed below, which have very different implications for cost and functionality, are ordered to illustrate increasing degrees of flexibility, with delivery of video used as the example of a core application of the network.

- Option 1—The undesirable option, which allows for no integration, is installed technology that is useful for only one purpose, in this case the delivery of video. Cable systems today are often this limited, offering no options for expansion of service. Future systems might also be designed for this single function if there is no broader vision of the range of services to be offered.
- Option 2—The system is designed to carry two separate services—multiple channels of video and the open bearer service. Some of today's cable systems have small additions (like bidirectional amplifiers) that allow a completely separate service such as the one Continental Cablevision and PSI are proposing to offer in Cambridge, Massachusetts. The two separate services cannot see each other, but the second could grow to be NII compliant in its own right.
- Option 3—The idea of separate services for video delivery and open bearer service becomes more interesting for future systems with digital video encoding. As the delivery systems migrate from NTSC encoding of video to digital encoding, such as MPEG-II, the video becomes more amenable to processing by both specialized and general digital processors. While the digital video may be delivered over the network in a highly specialized and cost-effective manner, the encoding can be defined (as MPEG-II is) so that it has at least one form that separates the actual video information from the details of the delivery method. A capability for converting the video into this format at customers' premises allows it to be transferred to and processed by other end-node elements such as general-purpose computers. This arrangement is a very powerful one: on the one hand, it does not affect the coding or delivery of the presumed high-volume information, which can be delivered in whatever cost-reduced manner the industry prefers; on the other hand, it simultaneously supports a general bearer service and provides a way to move the video into that more general format as needed at the end node. This report suggests that this combination be used as an example of a step that can be taken to realize the vision of an open and flexible NII.
- Option 4—Further options for integrating the delivery of video and the general bearer service would more directly couple the two, but

might also raise issues of cost and the utility of the network technology. One step would be to use the ODN bearer service as the delivery framing for cable TV. While some would argue that a properly defined bearer service could be used for a specific high-efficiency situation such as cable delivery of video, the uncertainty about cost impacts represents a significant issue that could impede this step. However, this step is not necessary in order to achieve an integrated ODN. Once the costs are understood, this degree of integration might make sense. This is a decision for industry to make.

- Option 5—Deploying the ODN could be further facilitated by designing an access network technology specifically intended to reduce the costs of the core service (in this instance, digital video delivery) and at the same time provide a carefully designed general, bidirectional ODN-style bearer service. The TV would ideally use an open bearer representation if costs permit, but this detail is not critical to the success of the effort. In the section "Research on the NII" at the end of this chapter, the committee notes a specific research and development effort that could be funded to move toward this goal.

The committee believes that a truly open NII cannot be achieved unless the cable TV and/or telephone access circuits are reconstructed in a way that supports the services of both the entertainment and information sectors. Technically, what is needed is a bearer service based on the concept of bidirectional digital packet switching. Such a service would permit the objectives of the ODN to be supported at the same time as other services dedicated to the entertainment sector.

Need for Government Action in Balancing Objectives

Because there will be real incremental costs in engineering access circuits to the home to meet broad NII objectives of the kind expressed in the Open Data Network,[41] the committee concludes that this is a time when the government, by direct action, can materially influence the course of the NII. At a minimum, the government could urge that future access circuits to homes be implemented in a way that supports the vision of an integrated NII, although, where real incremental costs are involved, simple urging will not be effective. At the other extreme, the government could mandate that all future access circuits support NII objectives, based on some specific set of service requirements. Such a mandatory action would preserve the concept of equitable cost for competitive providers, as occurs, for example, with various communications equipment features required by the Federal Communications Commission. However, in the current climate of limited regulation and increased dependence on competitive forces to shape industry, the tack of direct government mandate may not be feasible, and such precedents may not be followed.

The committee thus seeks a middle ground for government action, in which government actively works with the affected industries to define a suitable structure of economic incentives that serve to motivate the installation of a national infrastructure with appropriate characteristics. *The debate about next-generation television, entertainment networking, and the emerging reconstruction of the nation's access circuits has reached a crucial point: what follows may be either a departure from a broader NII, or a striving for interoperability with it. The committee strongly urges the latter approach, and it urges the government to recognize the opportunity at hand and to act on it.*

ACTING NOW TO REALIZE A UNIFIED NII

There is a definite role to be played by government in the pursuit of an open and flexible national information infrastructure, especially as commercial providers begin to figure ever more dominantly in the deployment of network technology. Left to their own devices, commercial providers will properly serve the markets that offer them growth and profitability. Planning for change, providing a general and open base-level service, and using and supporting open standards all may increase the cost of network deployment and thus may be at odds with commercial plans targeted to essentially one product, such as video delivery.

It is difficult to judge the degree of tension between the committee's vision of an open network and current commercial deployment plans. The true economic costs of openness cannot be judged by assessment of today's equipment, which was not designed to this end. The intrinsic support from the market for an open NII cannot yet be judged; this committee has assessed the experience in the limited communities (such as the academic research community) that have had real exposure to its vision but notes in those discussions the issues in scaling from those limited communities to a larger society. Finally, although it can discover anecdotal evidence that the tension exists, the committee necessarily has a limited ability to assess marketing plans and commercial projections.

The tension that is evident is more a question of degree than a choice between two poles. Nonetheless, it is a tension that must be recognized and rationalized if a coherent NII is to come into existence. In stating its approach, the committee recognizes that in the long run, it is the market, and not the force of government, that will determine if the vision of an open NII is relevant, useful, or successful. In attempting to identify those steps that could lead to the realization of an integrated NII, the committee notes at the same time the need to minimize the impact of guessing incorrectly. It is the committee's belief that, with proper technology development, taking the key steps now that will avoid precluding its vision can be done at tolerable costs. The committee urges the government to articulate a vision in this area and to work to ensure that industry will take this vision into account in its activities and development plans. Only if there is an accepted definition of NII compliance, and policy that encourages that compliance, will there be an effective way to accomplish a common vision of the future networks.

RECOMMENDATION: Technology Deployment

The committee recommends that the government work with the relevant industries, in particular the cable and telephone companies, to find suitable economic incentives so that the access circuits (connections to homes, schools, and so on) that will be reconstructed over the coming decade are engineered in ways that support the Open Data Network architecture.

The term "engineering" refers to the process of choosing what equipment to deploy, when to deploy it, and in what configuration to deploy it so that customer needs are met at least cost. The committee concludes that a national infrastructure capturing the ODN architecture will not be widely deployed if competitive forces alone shape the future; deregulation, along likely lines, will not be sufficient to guide the development and deployment of the ODN architecture. While anecdotal, numerous comments from inside the cable and telephone industries suggest that the perceived costs of adding the features that support openness are discouraging the necessary investment in the current competitive climate. The committee therefore concludes that these features will not be incorporated in the evolving national information infrastructure without policy intervention.

Needed now is direct action by government to ensure a planned, coordinated start to deploying the access circuit technology for the NII.

RESEARCH ON THE NII—ENSURING NECESSARY TECHNICAL DEVELOPMENT

This chapter touches on a number of areas in which continued research is required to realize the NII. The strong traditions of academic and industrial research that have led to U.S. leadership in telecommunications and networking must be continued and expanded if a truly national information infrastructure is to come into existence. Significant technical issues remain to be addressed in the development of the NII;

achieving the "Information Superhighway" involves more than just a matter of policy, legislation, and regulation.

It would be easy to conclude, from the great success of the Internet, that all required network research has been done. Such a conclusion would be very destructive to the leadership role played by U.S. industries and universities in all areas of technology related to information infrastructure. This committee believes that research, both experimental and basic, is essential to the future success of the NII and to national competitiveness. Chapter 6 summarizes the many ways in which the government can have an impact by continued involvement in the nurturing of research. In this section, the committee reviews specific technical areas in which work is required to fulfill the vision of an integrated NII. The section has been written to be self-contained and therefore repeats some concepts advanced in earlier sections.

For NSF to undertake support for the range of topics identified below would imply a significant expansion in the modes and paradigms for research that NSF has traditionally recognized. The list includes a number of specific research topics that would naturally fit into the traditional pattern of NSF proposals. It also identifies architecture studies and testbeds as research objectives. Architecture studies are often larger, more diffuse, collaborative, and less easy to define and to evaluate in advance. Testbeds, especially the virtual testbeds that have been built on the Internet, are again very collaborative efforts among workers at many sites, and they involve coordination and collective setting of direction as much as they do funding. Testbeds have been actively supported by ARPA, sometimes with NSF collaboration. A greater testbed effort by NSF would imply a departure from its normal pattern of funding, which involves the submission and evaluation of individual proposals from various sites and does not naturally lead to the required degree of direction setting, coordination, and architecture leadership.

Research to Develop Network Architecture

The Open Data Network architecture is a plan that defines the integrated NII's key aspects and how they fit together. The ODN architecture must be developed, a requirement that implies more than a program to study a series of technical issues. What is needed is a fitting together of all these pieces into an integrated concept that drives the whole development. This effort represents a research program in its own right and is perhaps the most important of the tasks that will lead to the NII.

The Internet is based on such a framework, which was first developed in the 1970s as an ARPA-funded research program. That research on architecture defined the key principles of the Internet: how the functions were divided into layers, how functional responsibilities were divided between the host and the network switches and routers, where conformance to a single standard would be mandatory, and so on. Similar architectural planning underlies any coherent infrastructure such as the telephone system, and it will definitely be required for the larger and more complex NII.

Network research today has tended to stress the issue of higher speed. The most obvious examples are the gigabit testbeds, jointly funded by ARPA and NSF, which have accelerated the deployment of high-speed network technology and stimulated the drive to higher-speed applications. However, the more pressing problems for networking tomorrow are issues of scale and heterogeneity, rather than speed. A system built so that it can scale to a large size is perhaps the most basic consequence of a successful ODN architecture. Problems of addressing and routing, of decentralization of management and operations, of dealing with heterogeneity, and of providing secure and trustworthy service are all issues that must be addressed in an overall architecture plan. *The government must support research into general and flexible architecture as a keystone of its NII research.*

Defining the Bearer Service

Definition of the ODN bearer service is an example of the sort of issue that arises as a part of architecture planning. One of the starting points for the ODN bearer service described by the committee was the IP protocol from the Internet protocol suite. However, as noted above in this chapter, even in the restricted domain of the Internet (as compared to the broader NII) the current IP services must be extended to meet emerging needs in areas such as explicit quality of service (QOS).

To address this issue, it is not sufficient to do research in how to implement QOS. The harder problem, now being addressed for IP in the relevant IETF working groups, is how to fit these concepts into the overall framework of assumptions that define the IP protocol. The most important architectural issue is to balance two objectives—to add flexibility to IP to make it more useful, and also to keep it simple and uniform—so that the protocol can be implemented in a consistent manner across a wide range of lower-level technologies. The power of IP is that it can be made to work over almost any network technology. Once the technology independence is augmented with complex requirements for QOS, this power may be lost. Balancing technology independence with the need for QOS is the essence of architecture research.

Presumably, the bearer service for the NII will have to reflect the integration of issues even broader than the issues associated with the next-generation IP. Developing the bearer service will require an understanding and balancing of a number of key technical factors. The outcome of this effort must be a scalable design that provides the needed isolation of the bitways below it from the information services above it; it must also be designed in a fashion such that a full range of QOS requirements can be met, including some that we cannot yet anticipate. *A directed research program toward this end would be a major contribution to the technical accomplishment of an open NII.*

Issues for the Lower Levels: Scale, Robustness, and Operations

The ODN architecture must, of course, incorporate a number of specific technical developments, each of which must be explored as part of the overall NII development. Summarized below are a number of issues, also raised early in this chapter, that are certainly under study today but are not yet sufficiently understood to enable meeting needs of the future.

Addressing and Routing

The issues of naming, addressing, and routing are among the most central to the success of a large-scale open NII. Both at the lower levels, where bits are being delivered, and at the higher levels, where services are being invoked, meaningful names must exist for the entities in question in order to make use of them. Without telephone numbers, one cannot call. Most successful network architectures, including the telephone system and the Internet, are struggling with the problem that the naming plan did not provide enough names to support the actual growth of the network.

Even more important, perhaps, is the problem of routing. As the network grows larger, and the range of services grows more complex, the difficulty of finding the location and a route to all the named objects gets more complex. This is a topic of much research at the present time, but there is yet no clear consensus as to the correct approach, taking into account all the real issues: decentralization of management, competing providers, mobility of end nodes (both computers and people), multicast delivery, worldwide scope, and so on.

Decentralization adds a substantial degree of complexity to routing. The various providers in the NII will each wish to make local assertions as to the sorts of traffic they will carry, and for whom. These various local decisions must be combined into a self-consistent route before any traffic can actually flow across the network. The problem of finding routes becomes even more complex when QOS is taken into account; the suitable route may depend on the details of the QOS specification, for example, a specific bandwidth requirement. When all these concerns are combined with the objective of establishing routes quickly, so that traffic is not delayed during route setup, the overall problem can be quite daunting.

There is a new generation of applications being deployed on the Internet, including audio, video, and shared work-space tools for multiparty teleconferencing, which depend for their operation on multicast (the ability to deliver information from a source to a set of recipients, instead of just a single recipient). Multicast makes the problem of finding good routes much more complex, since the range of options is greatly expanded. One approach is to build a separate route from each source to all its intended destinations. This approach is computationally complex but potentially the most efficient. Alternatively, one could build a single tree of routes that reaches from a known central point to all the destinations, and then allow any source to send to that central point as a way to accomplish multicast. This alternative is much easier to implement but is potentially highly inefficient, both in use of bandwidth and in extra network delay. Multicast is becoming a very important feature of the Internet, and research into its effective implementation is critical.

Even in the Internet there is concern today that the system has scaled to the point that the complexity of managing the routing at the global level will exceed current capabilities. Recognition of this problem helped to motivate the recent NSF award for routing-related research in connection with developing a routing arbiter function. A new generation of software tools will be needed to control and operate the even more complex NII.

Quality of Service

The NII will support a diversity of applications ranging from current Internet services such as e-mail, file transfer, and remote login to such new applications as interactive multimedia video conferencing, transmission of medical images, and real-time remote sensing. Each of these applications has different requirements for such QOS measures as delay, throughput, and reliability. Research is needed to understand how to design network components and interfaces that can provide the wide range of QOS that will be required in the NII. This work includes the design of specific mechanisms such as queue management and admission control, as well as the specification of general service models that the application can use to take advantage of these mechanisms.

New Approaches to Transport Protocols

TCP has been the default transport protocol of the Internet suite. However, it is not suited to all situations and all applications. TCP is designed to ensure totally reliable delivery. If it cannot deliver the information without error, it will not continue to transfer any information at all. However, some applications can tolerate errors. Audio and video for teleconferencing do not require totally error-free delivery. A momentary disruption in the transmitted stream is preferable to a total suspension of delivery. TCP as defined cannot provide this service.

Multicast, again, implies new issues for the transport layer. If one is sending to a set of receivers and an error occurs that disrupts transmission to only one of the receivers, the options for resolving the error are expanded. The transmission to all the destinations can be suspended until the error to the one is fixed, or the one receiver can be restored while communication to the others continues. There may be more relaxed models of reliability that make sense in certain applications, and there may be more options for restoring the state of the receiver.

Today, these cases are covered at the transport level by building into the application a special transport protocol that is custom designed for the application. There have been a number of proposals for a general framework for reliable multicast, but these have proved somewhat controversial, and no consensus has emerged. Research to generalize these ideas and propose a new generation of transport protocol would be very valuable.

Network Control Functions

Network controls are needed so that a very large and decentralized NII will be able to react to traffic overloads, network dynamics, and hardware and software failures. In addition to the routing function discussed above, a number of other network control functions will have an impact on the success of the NII. For example, when a user specifies a needed QOS, the network must decide if that request is one that should be accepted or not. If, for example, no route exists that can support the requested QOS, then the admission control function will reject the request.

On the other hand, a feasible route may be available that satisfies the QOS request, but the cost may be unacceptable to the user, in which case the request may be rescinded by the user; this is a form of self-imposed admission control, and it requires more interaction with the user at a high level. Once a call is accepted, the network has the responsibility of providing the agreed-upon QOS. In order to accomplish this, the network must "monitor" the traffic submitted by the user and must determine if that user is delivering no more than the level of traffic that was negotiated at the time of call acceptance. A number of congestion control methods exist that serve to guarantee that the contract is being kept by the user (one popular class being that of "leaky bucket control"), but no one technique has yet been universally accepted.

A proper study to determine an appropriate mix of these many control algorithms (routing control, admission control, cost control, congestion control, flow control, error control) has yet to be done, the outcome of which is necessary to set the proper guidelines for key aspects of the lower levels. All of these control issues are the subject of current research, many are being tested in the various networking testbeds, and work on these issues will support multiple needs such as those discussed under "Quality of Service" above.

Mobility as the Computing Paradigm of the Future

Mobility has to do with the capability to access data, resources, and services from any location via either wireline or wireless connections to the NII. One low-level issue in mobility is addressing and routing: how to deliver data to a host whose location changes as it is in operation. Another low-level issue is how and to what the network should assign names. A name could be assigned to the device that roams, to the user, to a location, to a user with a specific device at a specific location at a specific time, and so on. The higher-level issue, perhaps the more important in the long run, is how applications must change in the mobile environment, and what kind of network and operating system support is required to manage mobility. One implication of true mobility seems to be that communications may be intermittent. Medium- and long-term disruptions in the communications infrastructure will cause most applications today to fail, or at least to display to the user behavior that is quite unsuitable. New models are needed for information caching, coordination of local and remote versions, methods for providing services to mobile users according to their system profile, and so on.

Research is also needed to develop a system that can give a mobile user, who may be linked with a low-speed line to the infrastructure, feedback about how many bytes, dollars, or increments of time may be needed for the reply to a query made of the system. Such a capability would allow the user to either abort the request, or, in a system with the necessary intelligence, to ask for a subset of the reply, or an approximation. To date, only preliminary research has been done on the software architecture and related system interfaces that will enable mobility; this is a critically important area of investigation for the emerging NII.

Management Systems—Monitoring and Control

The NII will require continuous support to keep it up and running and to ensure that users have a high level of available service. Since the NII is likely to be a conglomeration of many heterogeneous interconnected networks, network management will be especially difficult. First, the management function will be distributed across the component networks, since each network operator will presumably be responsible for managing his or her own component of the overall system. Second, in a very large system the degree of trust and integration between the parts is likely to be reduced, which leads to greater problems in isolating and resolving problems. Research is needed to enable development of new technical means to alleviate some of these problems, for example, techniques based on new methods to assess current operating state and to share this state among neighbor regions.

Measurement and monitoring aid in understanding the functioning of a given network. As the network grows very large and very distributed, the collection of data, the bandwidth needed by monitoring tools, the requirements for storing and processing of the collected data, and so on are affected. Research is needed in support of better techniques for fault detection, performance diagnosis, and prediction, as well as automated or semi-automated measurement capabilities, reporting, and even repair.

New Technology for Access Circuits

As this chapter has noted, the design of access circuits (the "last-mile" technology) has much to do with the ability to provide a ubiquitous NII with the range of services envisioned by the committee. Picking among the design options is a matter of both available technology and policy.

Specifically, the government could sponsor research to explore the options for several key features:

- Access circuit technology that could provide very efficient delivery of identified traffic classes such as video and at the same time support the general sorts of services, including multimedia mixed data traffic, envisioned for the NII;
- Access circuits that provide cost-effective mixing of "bursty" traffic from several sources, as well as a means to deal with transient congestion by reflecting it back to the end nodes; and
- Access circuit technology that could provide for a cost-effective variation in services for bidirectional traffic, to accommodate a range of end-node needs for capacity into and out of the network. If the technology provided the ability to adjust this service dynamically, the investment

in installed infrastructure could better survive the changes in the pattern of usage that can be anticipated over the lifetime of the investment.

Middleware and Information Services Support

The previous section addressed some of the current concerns related to classical networking, the low-level transport of bits among end nodes. As the committee has observed, what distinguishes an information infrastructure from a basic data network is a defined and implemented set of higher-level services, the middleware layer, which provides an environment more directly suited to the advanced applications that will run there. The middleware layer is a much less mature component of networking than the lower layers, which have been explored and reduced to practice in a number of cases. Thus, the issues in middleware research are very open ended and wide-ranging.

Navigation and Filtering Tools

A major issue is how to tame the multimedia wave of information that is breaking over our heads: 500 satellite channels, terabytes of skyfall per day, thousands of new books and magazines and hundreds of thousands of newspapers and newsletters a day, and an exploding array of electronic mail and bulletin boards. How can people find the good stuff and filter out the rest? Much of this information is text or has a text component. Certainly that is true of print and mail, but it is also true of some video (if it is closed captioned) and documents scanned using optical character readers.

A major issue is how to "navigate" and "filter" text. Until recently, information retrieval has been a neglected subject of computer science and of library and information sciences. The work has tended to be very academic and has not focused much on either user interfaces or very large databases. Research is needed on text understanding, enhancing, and indexing and on filter and search interfaces and engines.

Nontext media (sound, speech, image, video) research is much more speculative. The problem statement is the same: capture, process, enhance, index, filter, search. But the task is much more formidable. We have no good techniques for categorizing such media (e.g., fingerprints have one category scheme, x-rays another, but most image classes do not have a well-defined categorization). Any search or index scheme requires such categorization.

Probably the major breakthroughs will come not from the research on search methods but rather from research on user interfaces. Today, there is a real barrier between people and machines. Graphical user interfaces are valuable, but the current interfaces between humans and our knowledge bases are very labor intensive because there are significant learning barriers to their use. They present a hypertext and keyword-search capability but generally do not incorporate contextual knowledge of the user. A breakthrough is needed in the research on user interfaces and query interfaces to facilitate access to query engines and knowledge bases.

Intellectual Property Rights

The NII promises a future in which the trade in information is as significant for the national economy as the trade in goods and services now is. Research is needed to make clear what new tools will be needed and how they will affect society. The solutions should promote the availability of a broad diversity of information over the networks, rather than simply control access. Some of this work was contemplated in recently proposed legislation; relevant work would also fall under the Information Infrastructure Technology and Applications (IITA) component of the HPCC initiative. Partnerships among industry, academia, and government will be essential.

Today we lack a consistent technical, legal, and business framework for the dissemination of intellectual property over networks. Problems to be solved range from licensing and fee recovery when material is being sold, to ensuring the integrity of information and proper attribution when the information is being freely distributed. If information owners have some assurance that licensing agreements entered into electronically over the network are enforceable, they will have significantly more incentive to trust information to this new environment. Information providers will also require some assurance that they are protected against liability, at least to the extent that they are protected when distributing their information over other types of electronic systems. Another capability that might tend to reduce unauthorized copying of intellectual property is an easy means for on-line payment of copying fees. Experimentation with technology that allows payment, in a simple and robust fashion, for services obtained through an electronic route is essential. The committee recognizes that such experimentation has begun in industry and university settings. Mechanisms based on both credit card charges and electronic money could provide attributed and anonymous purchase of information. A very efficient payment system would facilitate the selling of information in small increments, which is a natural transaction over a network. However, experiments with users will be required to determine if incremental or flat-fee payments are more desirable.

Computer and Communications Security

Classical end-node security is based on the idea that each node should separately defend itself by using controls on that machine. However, current-generation PCs and workstations are not engineered with a high degree of security assurance, so as a practical matter, an alternative is being deployed based on putting "firewalls" into the network, machines that separate the network into regions of more and less trust and regulate the traffic coming from the untrusted region. Firewalls raise a number of serious issues for the Internet protocol architecture, since they violate a basic assumption of the Internet, which is that two machines on the same internetwork can freely exchange packets. Firewalls explicitly restrict the sorts of packets that can be exchanged, which can cause a range of operational problems. Research with the goal of making firewalls work better—making them both more secure and more operationally robust—would be very important at the present time.

The strongly decentralized nature of the NII makes security issues more difficult, because it will be necessary to establish communication among a set of sites, each of which implements its own controls (e.g., user authentication) and is not willing to trust the other. Trustworthy interaction among mutually suspicious regions is a fundamental problem for which there are few general models. Security techniques using any form of encryption are no more robust than the methods used to distribute and store the encryption keys. Personal computers today offer no secure way to store keys, which severely imperils many security schemes. A proposal for solving this problem would be very important. The other issue with keys is the need for trustworthy distribution of keys in the large, decentralized NII. How can two sites with no common past history of shared trust exchange keys to begin communication in a way that cannot be observed or corrupted? The most direct solution would seem to be a trusted third party who "introduces" the two sites to each other, but there is no framework or model to define such a service or to reason about its robustness. Research in this area is a key aspect of fitting security into a system with the scale of the NII.

In addition to protection of host computers and the data they hold, the network itself must be protected from hostile attack that overloads the network, steals service, or otherwise renders the system useless. Additional research and development should be done on technical mechanisms, better approaches to operation, and new approaches to training and education.

Methods and technology for ensuring security are relevant to both the lower levels of the network and to the higher levels of the information infrastructure. Protecting intellectual property rights is a security concern, as is anticipating problems of fraud in payment schemes (control of fraud depends on identifying users in a trustworthy manner).

Again, achieving security requires the study of specific mechanisms and overall architecture. Much is known about techniques such as encryption. What is equally important is a proposal for an overall plan that combines all useful techniques into a consistent and effective approach to ensuring security. This overall plan must be developed, validated, and then replicated in such a way that users and providers can understand the issues and implications associated with their parts of the overall system. This effort, not the study of specific mechanisms, is the hard part, and the key to success.

Research in the Development of Software

The continuing need for research in means to develop large and complex software packages is not new, nor is it specific to networking and the information infrastructure. At the same time, it is a key issue for which there seems no ready solution. Problems of software development are a key impediment to realization of the NII.

A new generation of applications developed to deal with information and its use are likely to be substantially more complex than the application packages of today: they will deal with large quantities of information in heterogeneous formats, they will deal with distributed information and will be distributed themselves, they will provide a level of intelligence in processing the quantities of information present on the network, and they will be modular—capable of being reconfigured and reorganized to meet new and evolving user objectives.

These requirements represent a level of sophistication that is very difficult to accomplish with reliability, very expensive to undertake, and thus very risky. The committee adds its support to the continued attempts to advance this area.

Experimental Network Research

Experimental research, which involves the design of real protocols, the deployment of real systems, and the construction of testbeds and other experimental facilities, is a critical part of the research needed to build the NII. *Since this sort of work is often difficult to fund and execute, especially within the limits of the academic context, the committee stresses the importance of facilitating it.*

The Internet has provided an experimental environment for a number of practical research projects. In the early stages of the Internet, the network itself was viewed as experimental, and indeed these experiments played an important role in the Internet's development. However, the increasingly operational nature of the Internet has essentially precluded its use as a research vehicle. In the future, any remaining opportunity for large-scale network research will vanish, given that the NSFNET backbone is about to disappear and will be replaced by commercial networks and a backbone with only a small number of nodes, the very high speed backbone network service (vBNS), which is to provide high bandwidth for selected applications. In addition, it is likely that most of the Internet, like the larger NII, will be operated by commercial organizations over the next few years.

This transition has required the implementation of separate networks used specifically for research and experimentation. The gigabit testbeds provide facilities for investigating state-of-the-art advanced technologies and applications. ARPA has also provided a lower-speed experimental network, the DARTnet, connecting a number of ARPA-funded research sites. However, these networks are small and do not provide any real means to explore issues of scale. Indeed, there does not seem to be any affordable way to build a free-standing experimental network large enough to explore issues of scale, which is a real concern, since practical research in this area is key to the success of the NII.

Currently, the research community attempts to deal with this problem by using the resources of the Internet to realize a "virtual network" that researchers then use for large-scale experiments. Thus, multicast has been developed by means of a virtual network called the M-bone (multicast backbone) that runs over the Internet.[42] Similar schemes have been used to develop new Internet protocols.

There is a danger that the success of the Internet, much of which has been based on its openness to experimentation, will lead to a narrowing of opportunities for such experimentation. It is important that a portion of the NII remain open to controlled experiments. A balance must thus be maintained between the need to experiment and the need to provide stable service using commercial equipment. Attention should be given to the technical means to accomplish these goals. Funding should be allocated for the deployment of network experiments and prototype systems on the NII, even though they may be relatively more expensive than other research paradigms.

Experimental Research in Middleware and Application Services

Conducting testbed experimentation at the middleware level is usually less problematic than doing network research, because operation of experimental higher-level services cannot easily disrupt the ongoing operational use of the network by applications not depending on those services. The Internet thus remains a major facility for development and evaluation of middleware services, an opportunity that should be recognized and encouraged. Testbeds can address associated management of rights and responsibilities, including assessment of needs and mechanisms for the protection of privacy, security, and intellectual property rights.

Experimental and testbed efforts are needed to support a transition to higher-level, information management uses of networks. As John Diebold has observed,[43] applications of information technology progress through a cycle encompassing modernization of old ways, innovation (involving the development of new access tools and services), and ultimately transformation from one kind of activity to another (including doing the previously inconceivable). A great deal of experimentation is needed to achieve truly transformational applications.

The challenge can be illustrated by reference to the emergence of "casual publishing." The ability to publish from a desktop has changed publication practices; desktop video generation and reception will change them more. Although computer technology is making publishing changes possible, who can benefit, how, and at what costs will depend on the nature of the infrastructure. A similar set of technical, market, and policy issues arises in the digital library context, where experimentation has begun with support from NSF, ARPA, and NASA.[44]

Rights Management Testbed

More generally, an example of a useful testbed relating to rights management would incorporate systematic identification of the rationale for actions appropriate to government and industry into a joint industry-government project demonstrating model contractual and operational relationships to support the carriage of multimedia proprietary content. The computer, telecommunications carrier, cable provider, software provider, and content provider industries should participate, perhaps providing matching funds complementing a small contribution from the federal government, with broad dissemination of results a requirement. Questions that should be answered include the following:

- How can electronic authorization or execution of electronic contracts be provided over the network? This is an example of a general and flexible piece of infrastructure that the private sector is not likely to provide.
- What means can be developed to quickly provide varying degrees of authorization for particular uses of a work, for example, when the work may be used by different users for different purposes and at different pricing schemes?

- What various technological means—and the associated best times to use them—can be found for protecting data?
- What are the options for formatting multimedia information in a consumer-friendly fashion for distribution over the network to "episodic" users? This area is now the focus of considerable amounts of research and industry activity.

All efforts should be aimed at the most cost-efficient and interoperable means of achieving goals. A variant or a component of the above concept might include a series of multimedia projects that explore provision of electronic access to collections and materials generally inaccessible in the past, but of high research value, including photographs, drawings, films, archival data, sound recordings, spatial data, written manuscripts, and so on.

Research to Characterize Effects of Change

It will be important to understand how the evolving infrastructure will affect both the infrastructure for research and education as well as processes for research and education. This continuing process of change presents new challenges that militate against NSF assuming that it has successfully demonstrated the value of networking to research and therefore can diminish activity in that area. The new NSF-ARPA-NASA digital libraries initiative and NSF and ARPA information infrastructure-oriented activities under the IITA component of the HPCC program are steps in the right direction, but they are only first steps.

RECOMMENDATION: Network Research

The committee recommends that the National Science Foundation, along with the Advanced Research Projects Agency, other Department of Defense research agencies, the Department of Energy, and the National Aeronautics and Space Administration, continue and, in fact, expand a program of research in networks, with attention to emerging issues at the higher levels of an Open Data Network architecture (e.g., applications and information management), in addition to research at the lower levels of the architecture.

The technical issues associated with developing and deploying an NII are far from resolved. Research can contribute to architecture, to new concepts for network services, and to new principles and designs in key areas such as security, scale, heterogeneity, and evolvability. It is important to ensure that this country maintains its clear technical leadership and competitive advantage in information infrastructure and networking.

NOTES

1. The term "open" has been used in a variety of ways in the networking and standards community. Some of the uses describe rather different situations from that which is described in this chapter. For example, the telephone companies have been developing a concept they call open network architecture. That architecture does not address the concerns listed here; it is a means to allow third-party providers to develop and attach to existing telephone systems alternative versions of advanced services such as 800-number service.

2. Tolerance of heterogeneity must be provided for at more than the physical layer. At the higher levels, information must be coded to deal with a range of devices. For example, different video displays may have very different resolution: one may display a high-definition TV picture, while another may have a picture the size of a postage stamp. To deal with this either (1) the picture must be simultaneously transmitted with multiple codings, or the postage stamp display must possess the computational power of an HDTV, so that it can find within the high-resolution picture the limited information it needs, or (2) (preferably) the information stream must have been coded for heterogeneity: the data must have been organized so that each resolution display can easily find the portions relevant to it.

3. An illustration of this point can be seen in the history of a protocol suite called XNS that was developed by Xerox. XNS was proposed in the early 1980s and received considerable attention in the commercial community, since it was perceived as rather simple to implement. The interest in XNS continued until it became clear that Xerox did not intend to release the specification of one protocol in the XNS suite, Interpress, which was a protocol for printing documents. Within a very short time, all interest in XNS ceased, and it is essentially unknown today.

4. The notion of a multilayer approach is consistent with directions now being undertaken by ARPA and NSF in supporting the NREN and IITA components of the HPCC initiative. It also appears in such projects as the proposed industry-university "I-95" project to "facilitate the free-market purchase, sale and exchange of information services." See Tennenhouse, David, et al. 1993. *I-95: The Information Market*, MIT/LCS/TR-577. Massachusetts Institute of Technology, Cambridge, Mass., August.

5. The committee notes that conceptual models of the sort offered here may differ from models used to organize implementations, and emphasizes that the purpose of its conceptual model is to provide a framework for discussion and understanding. Models intended to guide actual implementation must be shaped by such issues as performance and may thus be organized in a somewhat different manner. In particular, a modularity based on strong layering may not be appropriate for organizing software modules.

6. This four-layer taxonomy is not inconsistent with a three-layer model that has been articulated in recent NII and HPCC presentations, a model based on "bitways," middleware, and applications. The taxonomy suggested in this report further divides the lower bitways layer to emphasize the importance of the bearer service, as is discussed in text below.

7. Quality of service (QOS) is discussed again later in this chapter. Although somewhat technical, this matter is a key aspect of defining the ODN. Today's Internet does not provide any variation in QOS; it provides a single sort of service often called "best effort." The telephone system also provides only one QOS, one designed for telephony. The Internet is currently undertaking to add user-selected QOS to its core service; it seems a requirement for a next-generation general service network.

8. In the Internet today, these transport features are provided by a protocol called the Transmission Control Protocol, or TCP, which is the most common version of the transport layer of the Internet. The TCP assigns sequence numbers to data being transferred across the network and uses those sequence numbers at the receiver to assure that all data is

received and that it is delivered in order. If a packet is lost or misordered, these sequence numbers detect that fact. To detect whether any of the data being transferred over the network become damaged due to transmission errors, TCP computes a "checksum" on the data and uses that checksum to discover any corruption. If a damaged packet is detected, the receiving TCP will ask the sending TCP to retransmit that packet. The TCP also contains an initial connection synchronization mechanism, based on the exchange of unique identifiers in packets, to bring an end-node connection into existence reliably.

While TCP is the most prevalent of the transport protocols used in the Internet, it is not mandatory, nor is it the only transport service. A range of situations, such as multicast delivery of data, and delivery where less than perfect reliability is required, imply the use of an alternative to TCP. For this reason, TCP is defined in such a way that no part of its implementation is inside the network. It is implemented in the end nodes, which means that replacing it with some other protocol does not require changes inside the network.

9. The transport layer defined in this report is not exactly the same as the layer with the same name in the OSI reference model, the OSI layer 4, because it also includes protocols for data formats, which are a part of the OSI presentation layer. Thus the ODN transport layer is a more inclusive collection of services that gathers together all the services that are provided in the networks of today to support applications in the end node.

10. For example, the government required that television sets include UHF tuners. In retrospect, most people would argue that the policy was seriously flawed. UHF television has never lived up to its expectations, the service has hoarded billions of dollars worth of valuable spectrum for decades, and the cost of television sets was increased with little net benefit to consumers—especially those living in less populated areas.

11. The committee recognizes that the Information Infrastructure Task Force has begun to explore the concept of technically based "road maps" for the NII.

12. The committee recognizes that unbundling is a controversial issue under current debate among state and federal regulatory agencies. The Ameritech proposals to open up its facilities present one indication that recognition of tendencies toward unbundling may be widening within industry. See Teece, David J. 1993. *Restructuring the U.S. Telecommunications Industry for Global Competitiveness: The Ameritech Program in Context*, University of California at Berkeley, April. This monograph describes how Ameritech offers to unbundle its local loops and provide immediate access to practically all local facilities and switching systems, with significantly lower costs for the unbundled loop compared to the revenue available from exchange telephone and related services:

> Once effectuated, the Ameritech unbundling plan will make the local exchange effectively contestable. Basically, anyone wanting to enter any segment could do so at relatively low cost. Entry barriers would in essence be eliminated. . . . [I]nterconnectors can literally isolate and either use or avoid any segment of the network. They are also free to interconnect using their own transport or purchasing transport from Ameritech. . . . [A]ll elements of the network must be correctly priced since any underpriced segment can be used separately from the balance of the network and overpriced segments can easily be avoided. (p. 64)

13. With each emerging network technology, including Ethernet, personal computers, high-speed LANs such as FDDI, and high-speed long-distance circuits, there have been predictions that IP or TCP would not be effective, might perform poorly, and would have to be replaced. So far these predictions have proved false. This concern is now being repeated in the context of network access to mobile end-node devices, such as PCs and other computers, and other new communications paradigms. It remains to be seen if there are real issues in these new situations, but the early experiments suggest that IP will indeed work in the mobile environment.

14. There are some other less central IP features, such as the means to deal with lower-level technology with different maximum packet sizes. There is also a small set of IP-level

control messages, which report to the end node various error conditions and network status, such as use of a bad address, relevant changes in network routing, or network failures.

15. A related issue is development of standard format sets for publishing over the Internet, for requiring headers and/or signatures, or for requiring some kind of registration that might automatically put the "work" in a directory.

16. Indeed, many applications cannot predict in advance what their bandwidth needs will be, as these depend very dynamically on the details of execution, for example, which actions are requested, which data are fetched, and so on.

17. Providing a refusal capability has implications for applications and user interfaces designed in the Internet tradition, which today do not ask permission to use the network but simply do so. The concept of refusal is missing.

18. By late 1993, perhaps 1,000 people worldwide were using real-time video over the Internet. The consequence was that at times fewer than 0.1 percent of Internet users consumed 10 percent of the backbone capacity. Personal communication, Stephen Wolff, National Science Foundation, December 20, 1993.

19. This point is relevant to a current debate in the technical community about whether the basic bearer service that can be built using the standards for ATM should support statistical sharing of bandwidth. Some proposals for ATM do not support best-effort service, but rather only services with guaranteed QOS parameters. This position is motivated by a set of speculations that a "better" quality of service better serves user needs. However, taking into account cost structures and the success of best-effort service on the Internet, a "better" service may not be more desirable. Technical decisions of this sort could have a major bearing on the success of ATM as a technology of choice for the NII.

20. The guarantee issue is related to the scheduling algorithm that packets see. A packet (or ATM) switch can have either a very simple or a rather complex scheduling algorithm for departing network traffic. The simple method is First In, First Out (FIFO). There are a number of more complex methods, one of which is Weighted Fair Queueing (WFQ).

In FIFO, a burst of packet traffic put into the network goes through immediately, staying in front of other later packets. WFQ services different packet classes in turn, so that the burst is fed into the network in a regulated way and then mixed by the scheduler with packets from other classes.

One alternative for achieving fairness is to allocate bandwidth in a very conservative manner (peak rate allocation) so that the user is externally limited (at the entry point of the net) to some rate, and then to assure that on every link of the network over which this flow of cells will pass, there is enough bandwidth to carry the full load of every user at once.

Such an approach using peak allocation eliminates from the network any benefit of statistical bandwidth sharing, which is one of the major benefits of packet switching. On the other hand, WFQ is one method for ensuring that we benefit from statistical multiplexing. The way this decision is settled will have real business consequences for the telephone companies and other ATM network providers.

21. One estimate for the accounting file for only long-haul intra-U.S. Internet traffic is that it would exceed 45 gigabytes per month per billion packets; the NSFNET backbone was approaching 40 billion packets per month by late 1993. See Roberts, Michael M. 1993. "Internet Accounting—Revisited," *EDUCOM Review*, November-December (December 6 e-mail).

22. This explicitly does not preclude implementing similar services in noncompliant ways as well. Thus, video might be provided according to the standards required for NII compliance, and as well in some proprietary noncompliant coding.

23. There is a great deal of uncertainty about the limitations of wireless. The low Earth orbiting satellites could provide considerable bandwidth; in this area, ARPA has funded a gigabit satellite experiment. At the local level (ground radio) the limitations are also un-

clear, but bandwidth will always be a problem to some extent with wireless. The question is how pervasive wireless will be for data communications. The predictions are indeed muddied. To quote a February 15, 1994, article in *America's NETWORK*, "The most experienced analyst with the best research data can't predict with certainty how the coming wireless data market will develop." Robert Pepper (FCC) reminds us that when cellular began, the best guesses were that there would be 1 million customers by the end of the century; today there are 60 million. It is clear that the lower-speed data services will surely be used widely. It is use of the high-speed services that is hard to predict.

24. The current status of the GOSIP is in doubt. NIST has convened the Federal Internetworking Requirements Panel to advise it on options for dealing with the GOSIP. At this writing, the draft report of this panel, opened for comments, was not yet final. However, the overall direction of the report appears to be to abandon the current GOSIP, which mandates one required protocol suite (the OSI suite), and to move to a more open approach based on multiple suites and an explicit acceptance of the Internet protocols.

25. See also U.S. Congress, Office of Technology Assessment. 1992. *Global Standards: Building Blocks of the Future.* TCT-512. Government Printing Office, Washington, D.C., March.

26. In the early 1970s, ARPA undertook the development of TCP/IP for the specific purpose of providing a standard approach to interoperation of DOD networks. The technical development was done by a working group convened and funded by ARPA, with academic and industrial research participants. In the late 1970s, ARPA worked with the Defense Communications Agency (DCA) to mandate TCP/IP as a preliminary standard for internetworked DOD systems. The DCA and ARPA cooperated on the establishment of a more formal review committee to oversee the establishment and deployment of TCP/IP within the DOD.

27. Additionally, the committee notes the emerging issues of addressing in the cable networks. Today, the cable networks have no real need for a global addressing architecture, since distinguishing between individual end nodes is needed only for directing the control messages sent to the set-top box. However, as the entertainment products become more complex and interactive, the need for an explicit addressing scheme will increase. If the cable networks expand to interwork with other parts of the information infrastructure, their addressing scheme should probably be unified with the scheme for telephony and information networks.

28. Computer Science and Telecommunications Board (CSTB), National Research Council. 1990. *Computers at Risk: Safe Computing in the Information Age.* National Academy Press, Washington, D.C.

29. Wireless radio transmission is especially subject to security risks for a number of reasons. First, the transmission is broadcast into the air, and so it is relatively simple to "tap" the transmission. Second, since the transmission is broadcast, a number of other radio receivers can easily receive it, and more than one of them may decode the message, opening up more opportunities for a breach of security. Third, since the medium is radio, it is easy to "jam" the transmission. Fourth, since radios are usually (though not always) portable, they are more vulnerable to being stolen, lost, damaged, and so on.

30. The committee recognizes that security is an emphasis of the administration's Information Infrastructure Task Force, but it seeks a sufficiently broad and deep technical framework, beginning with a security architecture.

31. CSTB has previously recommended more security-related research. See CSTB, 1990, *Computers at Risk.*

32. Lewis, Peter H. 1994. "Computer Security Experts See Pattern in Internet Break-ins," *New York Times*, February 11; and Burgess, John. 1994. "DOD Plan May Cut Ties to Internet," *Network World*, January 10, p. 95.

33. CSTB, 1990, *Computers at Risk.*

34. CSTB will launch a separate study of encryption and cryptography policy in mid-1994.

35. See CSTB, 1990, *Computers at Risk.*

36. Lewis, 1994, "Computer Security Experts See Pattern in Internet Break-ins."

37. Recognition of this problem is developing in the relevant industries, but problems of design and implementation remain. See "National Information Infrastructure and Grand Alliance HDTV System Operability," February 22, 1994.

38. Note that specificity is a theme of technology decisions for most interactive television trials to date. See Yoshida, Junko, and Terry Costlow. 1994. "Group Races Chip Makers to Set-top," *EE Times*, February 7 (electronic distribution), which observes that "many of the digital interactive TV trials and commercial rollouts are married to a particular set-top box design that is directly tied to a specific network architecture. Examples range from the set-top box Silicon Graphics is basing on its Indy workstation for Time Warner Cable's Full Service Network project in Orlando, Florida, to the box Scientific-Atlanta is building around 3DO Inc.'s graphics chip set for US West's trial in Omaha, Nebraska."

39. Continental Cablevision. 1994. "Continental Cablevision, PSI Launch Internet Service: First Commercial Internet Service Delivered via Cable Available Beginning Today in Cambridge, Massachusetts," News Release, March 8.

40. A wide range of speeds might be offered from the user back into the network. Today's options for access speeds range from voice-grade modems to current higher-speed modems at 56 kbps to ISDN at 128 kbps. None of these speeds are sufficient either for low-delay transfer of significant quantities of data or for delivery of video from the home into the network. Since high-quality compressed video seems to require between 1.5 and 4 Mbps, a channel of this size (at least a T1 channel) would permit a user to offer one video stream. It would still represent a real bottleneck for a site offering access to significant data. For comparison, in today's LAN environments 10 Mbps is considered minimal for access to data stored on file servers. Finally, for networks whose primary purpose is to provide access to entertainment video, the operator of the network presumably has the access capacity to deliver several hundred video streams into the network simultaneously. It is unlikely that this sort of inbound capacity will be readily available to any other user of the network. But at lower and more realistic input speeds, perhaps from T1 to 10 Mbps, there are a variety of interesting opportunities for becoming an information provider.

41. In an attempt to explore what these costs might be, the committee discovered that there are technical disagreements about the degree of additional complexity and cost implied by its objectives. Comments from inside the cable and telephone industries indicate that these industries have already assessed the costs of adding these more general features and have concluded internally that they cannot afford them in the current competitive climate. The committee thus takes as a given that these features will not be incorporated any time soon without policy intervention.

42. A virtual network such as the M-bone is constructed by attaching to the Internet a set of experimental routers. The operational IP addressing and routing is used to establish paths between these routers, which then use these paths as if they were point-to-point connections. Experimental routing algorithms, for example, can then be evaluated in these new routers. These new algorithms can neither see nor disrupt the operational routing running at the lower level, and so the experiment does not disrupt normal operation. The isolation is not perfect, however. In the case of the M-bone, quantities of multicast traffic might possibly flood the real Internet links, preventing service. Explicit steps have been taken in the experimental routers to prevent this occurrence. Building a virtual network requires care to prevent any chance of lower-level disruptions, since it does involve sending real data over the real network.

43. Diebold was quoted by Paul Evan Peters in a June 1993 briefing to the committee.

44. The NSF-ARPA-NASA digital libraries initiative solicits proposals for research in three areas: (1) "Capturing data (and descriptive information about such data) of all forms (text, images, sound, speech, etc.) and categorizing and organizing electronic information in a variety of formats." (2) "Advanced software and algorithms for browsing, searching, filtering, abstracting, summarizing and combining large volumes of data, imagery, and all kinds of information." (3) "The utilization of networked databases distributed around the nation and around the world."

Examples of relevant research are listed in National Science Foundation. 1993. "Research on Digital Libraries: Announcement," NSF 93-141. National Science Foundation, Washington, D.C.

3

Research, Education, and Libraries

The research, education, and library communities have been information infrastructure pioneers. Their use of networking to create new channels for scholarly communication and collaboration is pointing the way to broader and deeper participation by individuals at all levels of society in the process of learning. The Internet has been the vehicle for most of the networking explorations of research, education, and libraries, but it so far has been used most heavily by segments of the research community.

"Research" includes scientists across a wide range of disciplines as well as research in the humanities; "education" includes K-12 and a range of higher- and continuing-education institutions and activities; and "libraries" includes research and specialty libraries typically associated with research institutions as well as public libraries that, like K-12 schools, are embedded in a general community setting.[1] Research, higher education, and libraries subsume a number of elements that are sometimes separately addressed under the health care umbrella, such as health-related research, education, and library support.[2]

The experiences of the research, education, and library communities illustrate some of the tensions among what different groups of communities need, want, and can afford—tensions that will be replicated or expanded as the nation moves to a truly national information infrastructure (NII). They also illustrate that the value of the Internet—regardless of whether that value is measured in terms of program effectiveness, productivity gains, or returns on investments—increases as a function of (1) the size and diversity of its user population, (2) the power and sophistication of its applications, and (3) the capability of the infrastructure.[3]

Movement toward an NII and away from stand-alone, separate research networks will solve some problems but may give rise to others for the research, education, and library communities. To understand and provide adequately for their future networking needs, it is necessary to examine how these communities differ from other communities to be served by the NII, and to appreciate as well their continuing force for positive change in the growth of U.S. society.

RESEARCH

Within the research community, individuals and disciplines differ in terms of their use of electronic networks: some primarily use electronic mail, some emphasize access to shared databases, some require access to special tools or devices (e.g., supercomputers and special sensors), some depend on networks for rapid transfer of large data files, and some take advantage of "news" (i.e., discussions and bulletin board groups), as there are thousands of "newsgroups" addressing a multitude of topics and disciplines. The upgrade of the NSFNET backbone to T3 service has enabled new applications such as teleconferencing, multimedia electronic mail, visualization, and on-demand electronic publishing (Box 3.1). Qualitative benefits have arisen with changes in the nature of the work being done: broader interaction can change the questions being asked, the review accorded to research, and the scope of participation.[4]

Quantitative benefits or gains in efficiency may be seen in the sharing of scarce resources (obviating the need for duplicative investments), from the network facilities themselves to expensive devices and information resources attached to them. Another quantitative benefit is the effective reduction of the cost of reproducing and distributing information—it is often cheaper to duplicate and transmit information electronically rather than in paper or other physical media.

For example, a University of Colorado group funded by the National Aeronautics and Space Administration (NASA) built a prototype system for distributing satellite data over the network via an on-line access system. In this system the user is responsible for locating, reviewing, and ordering the data, an approach that saves human time and cost in locating, copying, and shipping data tapes. The system is made possible by a network that can accommodate the transfer (FTP) of modest-size data files (the size of a typical file is 1 to 4 Mbytes), using the usual techniques of data compression and user selection of data file size to minimize the impact on the network. During the recent TOGA-COARE[5] oceanographic experiments in the western Pacific, satellite sea-surface-temperature maps were produced and made available to users over the Internet; researchers not having Internet access required production of a contour

Box 3.1 Research Applications of Networking

- Communication via electronic mail with other researchers locally, nationally, and worldwide
- Sharing and transfer of data files among researchers, and between researchers and data sources (especially government agencies)
- Contribution, sharing, and accessing of news information of all kinds (e.g., conference announcements, current affairs, electronic bulletin boards, meeting abstracts, developments in individual fields, and so on)
- Electronic reference searches (e.g., access to on-line library catalogues, directories of special collections, and databases with abstracts)
- Access to special-purpose computing resources, such as supercomputers and sensor-based instrumentation
- Access to shared and community resources, including national and global databases and data systems
- Electronic submission of reviews and proposals
- Access to articles, books, and other materials published electronically
- Development of and access to community software
- Access to remote, remotely controlled instruments for research

map of the images that could then be faxed to them. Even the ability to electronically access text can present real savings: physicists, for example, have enthusiastically developed and used electronic archives in areas such as high-energy physics that are stored at Los Alamos National Laboratory facilities as an alternative to buying journals that cost hundreds to thousands of dollars per year for subscriptions.

One of the many new Internet tools that has facilitated information sharing and collaboration in biomedical research is the Mosaic hypermedia browser used by researchers studying the genome of the fruit fly, *Drosophila melanogaster*. Through the efforts of a large number of researchers in many laboratories, approximately 90 percent of the genome in this species has now been cloned onto 10,000 fragments, where approximately 3,000 conventional loci have also been mapped. A researcher at any participating site can open a computer screen window, use a mouse to click on (select) increasingly detailed photographs of the region in which he or she is interested, and then open lists of the relevant clones, deletions, and loci of that region, which include the corresponding literature references. This software was first developed for the nematode *Caenorhabditis elegans*[6] and is now available for a variety of other species.

Laboratories throughout the world, connected by the Internet, have begun collaborating to sequence completely the DNA of several organisms, leading ultimately to sequencing of the entire human genome.[7] The databases that must be linked include DNA sequences (GenBank), chromosome mapping information (Genome Data Bank), and protein sequences and structure (Protein Information Resource).[8]

New network capabilities and technologies will generate a continuing flow of new network applications that will lead to significant changes in the conduct of research. If development of high-performance computing and communications technologies in academia and industry continues, network capabilities should expand at rates that may initially exceed the demand generated by the science community. The increase in capabilities will make possible an overall shift from electronic mail exchange and modest file transfer activities to on-line distribution of large volumes of data (both measured and modeled), fueling in turn continued rapid growth in scientific researchers' demands for network services. One possibility is that high-performance computing in the future will involve networking dozens to thousands of workstations and other computers in different locations. Initially these computers will be running applications in the background and at night; but at some future time, people may collaborate through relatively continuous collaborative computing. Given such possibilities, it is hard to predict the ratio of transactions to bandwidth over time. *It is clear that the "experiment" of network use in the research environment is only just beginning.*

The nature of research is itself changing. Today scientists are seeking answers to more complex problems, while the instruments and facilities needed to conduct research are becoming increasingly expensive and the funding for scientists' projects is becoming scarce.[9] This is a scenario that strongly encourages increased collaboration, although the nature and extent of collaborative research may differ within disciplines and across disciplinary boundaries. There is considerable impetus for collaboration in oceanographic research, for example, where research projects not only cross the lines of the traditional subdisciplines of the field (i.e., physical, biological, chemical, and geological) but also require collaboration among international scientists.[10]

Aware of such trends in funding and the increasingly interdisciplinary nature of such research, some scientists have recently introduced a new concept: a center without walls in which the nation's researchers can perform their research without regard to geographical location—interacting with colleagues, accessing instrumentation, sharing data and computational resources, and accessing information in digital libraries. The name given to this concept is "collaboratory," a term derived by combining the words "collaboration" and "laboratory."[11] In this environment a scientist's instrumentation and information are virtually local, regardless of their actual locations; research teams separated by continents and disciplinary boundaries will be able to conduct joint experiments in ways that will greatly expedite the transfer of knowledge and thus will change the scientific process involving interdisciplinary and collaborative research. The collaboratory concept suggests that in addition to transmission and switching capability, information infrastructure for research will also need to provide more basic services and tools, such as mechanisms to easily identify and assemble the needed experimental components from across the network.

Today's research projects are increasingly multisite, and many involve numerous entities; universities, laboratories, industry, and other organizations may participate. Principal investigators are often geographically distributed, but they require opportunities for real-time interaction and, in many fields, capability for remote experimentation. For example, research in quantum chromodynamics, involving efforts to investigate and define intranuclear forces, is heavily data oriented and requires large amounts of bandwidth. Moreover, applications consume and produce huge amounts of data that need to be stored, processed, and transferred at different sites. To meet such a range of needs, a network service must support multimedia (including real-time audio and video) traffic, provide large amounts of bandwidth to those sites requiring it (smaller amounts to those that do not require the "kitchen sink"), and be quasi-ubiquitous in order to connect investigators and sites. Such a network service is also needed to provide for others, such as educators, access to the eventual results of scientific research and to enable educators to carry out their own research using network resources.

An example of the emergence of international, collaborative research is the International Thermonuclear Experimental Reactor (ITER), a Department of Energy (DOE)-sponsored project involving a multinational team (from the United States, Germany, Russia, and Japan) that is trying to design and build a reactor for energy production. The collaboration is among four major sites (one in each of the four countries) with additional other sites participating both within the above-named countries and elsewhere. The project has progressed to the stage that participants need the ability to exchange engineering (computer-aided design/computer-aided manufacturing) designs in real time and to concurrently analyze, discuss, and annotate these designs. Such capabilities require high-bandwidth connections for the transfer of engineering data and for the support of a "real-time" collaborative, distributed (internationally) environment. Today's networks do not have such connections. Near-term improvements may entail increasing the capacity of the deployed infrastructure; greater support for real-time collaboration would entail extending the infrastructure architecture along the lines outlined for the Open Data Network described in Chapter 2.

Several groups of researchers are already beginning to move large volumes of scientific information over the Internet. Those conducting global change research (oceanographers, earth scientists, and atmospheric scientists, among others), for example, obtain massive quantities of information from satellites and other collection devices, and those quantities are expected to increase to 1 terabyte per day through the implementation of the Earth Observing System Data and Information System (EOSDIS) and other programs organized through NASA, DOE, the National Oceanic and Atmospheric Administration (NOAA), and other federal agencies. The contemplated data volumes are so large that even with substantial (e.g., factors measured in hundreds) amounts of data compression, high levels of bandwidth will be needed to do research in these areas.

For example, researchers investigating global climate change currently use models that require very large amounts of data before, during, and after construction of the models. Their analyses are typically done visually with a high-end workstation. The limiting bandwidth available on ESnet and other segments of the Internet constrains the ability of these scientists to view their results in real time and, even worse, prohibits them from performing an analysis that incorporates both surface and air models—since these models are implemented on different machines in different locations, and very large amounts of data would have to be transferred for use in an integrated model. An additional complication for global climate change researchers is that data collected in NASA's Earth Observing System (EOS) is also likely to be used. Because these data will be located at various sites, including Oak Ridge National Laboratory, the transfer of large volumes of information will be required.

Similar requirements can be seen as arising from the growth in applications involving graphical images and video. One driver for these applications is scientific visualization, which makes massive data sets more comprehensible. High bandwidth will also continue to be required to support real-time interactions (e.g., to support remote access to special instruments or even video conferencing) and remote access to high-performance computing devices—the performance of which continues to increase.[12] In addition, interactive activities such as commanding spacecraft and monitoring remote instruments entail networking needs that greatly exceed current capacities and sophistication.

For example, the Advanced Photon Source (APS) comprises 22 separate collaborative action teams that involve industrial companies, national laboratories, and universities accessing the 40 to 50 APS beam lines. These teams require the ability to employ "telework" tools and techniques to perform remote experimentation and collaboration in real time. The hope is to extend this ability to an educational setting in at least a real-time monitoring or visualization mode. That prospect cannot be realized until the network supports multimedia traffic, provides the needed large bandwidths, and provides a more secure environment.

A central question regarding the supply of information infrastructure to the research community is the distribution of needs. At this time, specific communities can be identified that need very high bandwidth. The expectation that this group is relatively small is reflected in the NSF's plan to support a very small research network operating at very high speeds (the vBNS). How large will requirements for such infrastructure grow, and how quickly will they spread? The answers may depend on the degree of ease of access to high-bandwidth infrastructure.[13]

The degree of innovation, the diversity of applications, and the dependence of researchers on the Internet have been nurtured by an environment in which cost has not been a major constraint for the individual user (see Chapter 5). The benefits of connection to the Internet are underscored by contrasts within the research community: researchers in disciplines and at campuses without easy (or any) Internet access have not had the opportunities to collaborate, learn new results quickly, and broaden their professional interactions that their connected colleagues have.

The problems and opportunities inherent in broadening access within the research community are especially evident outside the natural and physical sciences. New technologies and Internet access are opening up new avenues for humanities research, teaching, and scholarly communication. Used far less commonly but enthusiastically among a small group of proponents, humanities computing and networking have grown significantly in recent years. Electronic bulletin boards, e-mail, and discussion groups are the most widely used mechanisms for day-to-day communication. In addition, literary text analysis can be carried out with unprecedented detail due to the availability of machine-readable texts and complex text-analysis software.[14] Thus, the Internet has allowed humanities professors at several institutions to assist in building a multimedia database for the study of ancient Greece under the Perseus project. With the collection and storage of text, translations, color images, maps, and drawings contributed by museums and archaeologists, the Perseus database will bring the world of ancient Greece alive for students and researchers across the country.

Humanities buildings, departments, and scholars are often among the last on a campus to be connected to the Internet.[15] Disciplines in the humanities have not been well capitalized in part because the need for capital (such as computing equipment and networks) has not been recognized, yet without access to the capital these researchers cannot demonstrate its value in their fields. Overall differences in funding across disciplines reflect societal choices about where to invest. The new National Initiative on Arts and Humanities Computing is currently planning the next steps in gaining a voice for the humanities and arts in the development of the NII.[16]

Broadening of access was a goal of the original National Research and Education Network (NREN) proposals, and substantial progress has been made through expansion of the original NSFNET, the Internet, the National Science Foundation (NSF)-catalyzed regional networks, and the NSF Connections program, as well as growth in the targeted ESnet and NASA Science Internet (NSI) programs.[17] The Branscomb report[18] has called for broadening the access to advanced workstations across the scientific research community, a move that would fuel demand for network-based infrastructure. The possibility of slower growth in research networking—implied by a shift toward commercialization and user payments compounded by some tightening in the availability of research support—raises the prospect of competition for infrastructure resources between those with no access (who argue for ubiquity and aggregate bandwidth to serve them) and those with high individual needs for bandwidth. However, as has been seen in the provision of access to ESnet and NSI, there may be special solutions developed to meet special needs. There is little question that researchers in engineering, science, health, and the humanities have benefited greatly from their use of the Internet. That benefit can be expected to grow considerably with the greater reach and capabilities of the NII.

HIGHER EDUCATION

Higher education networking presents a chicken-and-egg problem. Networking for this and other segments of the education community attracted federal attention somewhat later than networking for research partly because teaching faculty, most college students, and academic librarians—except for those involved with scientific research—had no access to the network until it was opened to them in 1988 with the expansion of the original National Research Network program into the NREN program. What they had were only their own visions of potentialities, often growing out of experiences with television and satellite broadcasting, which introduced telecommunications into the teaching-learning process.[19] The renaming of the federal NREN program to include education was a result of the library, academic computing, and education communities' intense, aggressive lobbying, political activity that also spurred the development of new organizations and activities dedicated to developing and applying information infrastructure for higher education.[20]

For example, the push for the "E" in NREN was associated with the establishment of the Coalition for Networked Information by the Association of Research Libraries, CAUSE, and EDUCOM.

On the whole, higher education is positioned to make extensive use of the NII to enhance the delivery of academic programs and to enrich their content by broadening the mix of inputs and participants. For example, BESTNET is an international consortium of colleges and universities using the capabilities of telecommunications to enhance the teaching and learning experiences of students across geographic and cultural boundaries. Using computer conferencing, e-mail, shared databases, and interactive computing, participating institutions are experimenting with collaborative teaching. Initially started as a cooperative effort among three California State University campuses, Texas A&I University (Kingsville), the Centro de Enseñanza Técnica Y Superior (Mexicali via Baja California Norte, Mexico), and Tijuana Instituto Technologico de Mexicali to share Spanish language instruction, it is now expanding its reach into other continents and other disciplines.

The NII will also help educational institutions to reach students where they are or prefer to be located. Students, especially graduate students, will want to pursue degree programs and professional development courses at teleconferencing sites that are convenient for them. For many students, especially older or adult students, convenient teleconferencing and remote access, in combination with the quality of their courses, may become more important than many of the campus experiences that accompany the traditional attending of classes at the institution itself.[21] In addition, the "electronic outreach" capability provided by the NII can enhance the recruitment and retention efforts of all educational institutions.

X Window and the Mosaic hypertext interface for the World-Wide Web offer broad possibilities for applications in undergraduate education. For example, Mosaic is incorporated into some of the "electronic studios" being developed at Rice University for courses ranging from an introductory biology laboratory to a graduate seminar in architecture. Electronic studios combine a variety of computing tools, some available over electronic networks. Tool sets vary by discipline; engineers may need circuit design programs, for example, while sociologists may seek statistics programs.[22]

The use of interactive video as a medium for the delivery of instruction in multiple sites is beginning to emerge. For example, the "CU-SeeMe" videoconferencing program was used in the Virtual Design Studio project for collaborative housing design among Cornell University, MIT, the University of British Columbia at Vancouver, the University of Hong Kong, Washington University at St. Louis, and an institution in

Barcelona, Spain. Three of the teams worked for two weeks using CU-SeeMe and gave a final video presentation using a PictureTel videoconferencing system. According to Kent Hubbell of the Cornell University Architecture Department, teams designing with CU-SeeMe were far ahead of the teams that did not use the technology: "It is really amazing what a difference just a little live video image makes to the entire communication process."[23]

Shared databases for print, video, voice, and image are anticipated. The cost savings possible from resource sharing are illuminated by existing information resource applications within higher education. For example, an increasing number of colleges and universities are contracting for networked access to information resources available from commercial providers. The California State University (CSU) system, with 20 campuses serving 350,000 students, illustrates the potential for significant savings on the acquisition of information resources through partnerships with vendors, such as Mead Data Central and Dow Jones News/Retrieval, which have been willing to offer services at a discount in return for being able to effectively promote their services to students. For example, the database service set known as LEXIS-NEXIS is accessed by students and faculty at all CSU campuses (at present, by approximately 200 concurrent users system-wide) through the Internet. At the average commercial rate of $10 per search and at the current volume of 125,000 searches per month by CSU users, the service would cost $1.25 million per month and the total annual fee, at commercial rates, would amount to $15 million; under its negotiated agreement with Mead Data Central, CSU pays approximately $200,000 per year. Note that while CSU was able to use its centralized procurement to its advantage, many libraries and educational institutions have raised concerns about the cost of site licenses relative to the service demand they generate.

The principal constraint on network-based applications is one of access; significant numbers of higher-education institutions remain with limited or no Internet access. By early 1994, approximately 1,100 institutions of higher-education (including all schools in the top two Carnegie classification categories, "Research" and "Doctorate") were connected to the Internet; the total number of higher-education institutions in the United States exceeds 3,000. The limited extension of the intracampus telecommunication infrastructure into classrooms and faculty offices, the shortage of desktop computers, the lack of resources for faculty development, and the scarcity of technical staff support have been major obstacles in programs mounted to date. Without extensive stimulation at state or federal levels, it may be more than several years before there will be ubiquitous access at institutions not already connected to the Internet.

One illustration of the challenge is provided by conditions in Michigan. Michigan-based MichNet was founded (circa 1970) to provide data networking connectivity among Michigan's publicly funded universities. Merit, the service provider, began by interconnecting the three largest public universities, i.e., Michigan State University, the University of Michigan, and Wayne State University, in 1971. Today MichNet provides 135 dedicated attachments to 92 organizations distributed over all of Michigan's two peninsulas. These include 42 four-year colleges and universities (including all 13 public universities) and 11 (of 29) community colleges.[24] Thus, only approximately one-third of Michigan's community colleges are served by MichNet currently. Because there is a statewide network backbone in place in Michigan, there is not a technical barrier to attaching new organizations. Rather, for each category of organization MichNet serves, the two primary inhibitors are funding and motivation. Community colleges, for example, like other organizations, usually weigh committing funds for data networking against other communication priorities, such as distance learning (which can be supported with conventional telecommunications). With the current attention being given to the Internet and the NII, there is increased awareness and interest in data networking access. But funding remains a challenge, although some relief has been found from NSF's Connections program.

K-12 EDUCATION

This is a time of great opportunity to expand information infrastructure for K-12 education, because such applications have attracted considerable political attention and support. The expansion of programs relating to mathematics and science education at NSF, the launch and proposed expansion of the National Telecommunications and Information Administration (NTIA) Telecommunications and Information Infrastructure Assistance Program, the signs of greater interest in technology at the Department of Education and that department's inclusion in the new Information Infrastructure Technology and Applications (IITA) component of the High Performance Computing and Communications (HPCC) program, the administration's challenge goal of connecting all classrooms to the NII by the year 2000, and industry efforts to connect schools make this a time of great expectations.

However, the K-12 education community to date has not generally been a major consumer of information technology. Some infrastructure applications are already in place, one example being the distance learning techniques that were rapidly deployed and accepted at many institutions, although these efforts may rely on more conventional telecommunications or other media. While networking has spread in the research community through the relative ease of access to seemingly "free," shared networks and the "contagion" effects of peer pressure or observation of colleagues, the number of educators with direct experience in using this technology has been limited to the early adopters, those who had special funding, or those with access to special single-purpose administrative networks.

Networking for education has followed a significantly different history from research networking. First, where it has occurred it has been inherently more expensive, characterized by the prominence of dial-up access (at telephone service rates, plus the applicable computer service charges, usually including on-line connection time charges plus occasional transmission charges) to bulletin-board or other central-host-based services. Much of this activity has been independent of the Internet, at least until relatively recently; networking in K-12 education has been inhibited by the lack of broad access to an open information network. There have been multiple services serving small and separate communities (militating against the achievement of a critical mass of users on any one service). Cost has been a major obstacle, given the chronic budget pressures confronting schools and the lack of predictable costs associated with educational networking.

Indeed, cost is a key reason that data networking connectivity among the educational community has basically only trickled down from the major research universities to other four-year schools and then to the two-year schools, with the K-12 and public library communities bringing up the rear. For example, Michigan has over 5,000 K-12 school buildings, a number dwarfing the 135 MichNet attachments (for all kinds of research and education institutions) in place today. To bring the Internet to all these sites will be a major challenge. As a result of an Ameritech overcharge, Michigan's K-12 community has a one-time opportunity to access additional funding to enhance its networking access. These funds may be used in various ways, for example, for video conferencing, distance learning, and data access. Yet even if all the available funding were used for Internet access, at best about 25 percent of the schools would have some of their access costs covered.[25]

Second, networking of any kind has been difficult to launch in education because of a lack of basic physical infrastructure. *The majority of K-12 educational institutions are ill-equipped to participate in an information society.* At the most basic level, schools often lack telephone connections and even sufficient electric outlets to support broad—or sometimes any—computer-based access to electronic networks. The deployment of telephones in the schools was limited because telephones were viewed as business tools, not instructional tools. By contrast, television technology was adopted quickly because the connection to instruction was easier to understand. In the business plans of cable franchises, many cable companies offered to network the schools as an incentive for the local community to accept their offer. Before long, school hookups were a standard part of the franchise agreement.

Although research users of the Internet typically effect connections through local area networks, where LANs are found in K-12 schools, the motivation has often been to support computer-aided instruction in a single classroom rather than to link multiple classrooms or provide access to external networks. Also, many personal computers in schools are not capable of connection to LANs and/or they may not have the necessary disk or memory capacity to support network applications.[26] Finally, schools also lack the resources to install new technology.

Upgrading computing and communications within schools is complicated by what are often inefficient approaches to procurement: procurement on an individual school or even a school district basis may not provide the volume sufficient to gain meaningful discounts. This situation suggests that even within specific areas (such as states), it should be possible to organize procurement to achieve economies. More efficient procurement can better leverage resources provided through corporate activities, such as programs like a recent Pacific Bell proposal to connect schools in its service area.

Third, a critical infrastructure deficiency relates to the human element: training and support in electronic communication are needed for educators. Most efforts to engage in networking for education have been undertaken on a do-it-yourself basis, requiring substantial dedication and time to overcome the difficulties of installing, using, and maintaining sometimes inscrutable systems. Usually, the first teacher who has experimented with the technology is labeled the "computer expert" and is rewarded by being given more responsibilities to cope with during a very busy school year. A large number of educators need to be trained to use the new methods of communication. This problem particularly affects schools serving lower socioeconomic students, where educators generally have less training and awareness of the Internet (and therefore are less likely to even desire access). The need for professional development in network applications and resources includes both the classroom instructor as well as the building-level administrator who will adopt the new technology.

Part of the problem is a lack of involvement on the part of teacher education programs and institutions. Colleges of education have tended to have limited network access. Teachers emerging from preparatory programs that have used electronic networking tend to ask for access, but most such programs provide little such exposure.

Of the many educational applications and benefits foreseen (Box 3.2),

Box 3.2 Educational Applications and Benefits of Networking (via Internet)

- Access to more current information (e.g., space science facts, weather patterns, White House press releases, and so on from national laboratories, government agencies, universities, and commercial sources) for use in developing curricula, assignments, and so on, and for increasing motivation in both students and teachers
- Access to more accurate factual information, in the social as well as natural and physical sciences
- Familiarization of teachers, administrators, and students with computing and communications technologies, for both educational and job-preparation benefits
- Experimentation: capacity to assess how networking fits into the curriculum, or if it does not fit, to determine how the curriculum might be reconceived
- Development of collaborations among students, teachers, and school administrators, building on common interests and experiences despite differences in location and strengthening a sense of belonging to one or more communities
- Ability to enable more active (as opposed to passive) acquisition of information and learning, increasing the interaction component of the educational process and facilitating a shift from secondary to more primary sources of information
- Reinforcement of basic skills of reading, writing, locating information, and structuring and solving problems
- Expansion of interest in science, through use of information resources provided by federal science agencies, sharing of other materials, and communication with others nationally on related topics (e.g., environmental protection, shuttle launches, geography, languages, and so on)
- Ability to follow up on professional development activities for teachers and administrators
- Ability to build a bridge from school to home through network links for parents and guardians, providing information about assignments, school events, curriculum content and structure, and so on

communication among people is the prime benefit of networking for education as colleagues, teachers, students, and parents gain access to a level of communication not possible before. Educators in both higher education and K-12 education are now being enabled to participate in collaborative projects. Much like the science "collaboratories," the Induction Year project at the Regional Education Source Center in Huntsville, Texas, which has provided an electronic support network for first-year teachers with mentors both in the public school and among university faculty, suggests new avenues for collaboration. A variety of international collaborations have developed among educators connected to the Internet.

As educators do gain access to the Internet (and other networks), their applications are evolving. As a result, it is dangerous to assume that educational networking needs are inherently simple and "low-end." For example, K-12 programs are using scientific data generated by such agencies as the DOE, NASA, and NOAA; there are K-12 programs (e.g., SuperQuest) providing access to high-performance computers; and graphics, video, and multimedia programs are inherently attractive to educators trying to convey complex information to children and other students. Nevertheless, given the constraints outlined above, much educational networking to date has been low-end.

To achieve many of the benefits anticipated by educators will require access to the high-end networking that would make possible better video and multimedia exchanges. This implies higher bandwidth, reliable service, and so on. More sophisticated systems and higher bandwidth enable better graphical interfaces and functionalities, which can reduce training costs, possibly offsetting higher transmission costs.[27] Use of the latest Internet tools such as the World-Wide Web and its Mosaic interface are limited only by the bandwidth and hardware found in K-12 institutions. Other important ingredients include development of suitable content and curricula. Networking in K-12 will never achieve critical status unless mainstream educational services are available over the network.

Perhaps most importantly, information infrastructure holds the promise of changing the processes of teaching, education, and learning (Box 3.3). Where this happens, resources will be reallocated from print-based to on-line resources (depending, in part, on publishers' support for electronic source material). Educators' roles are changing to include participation as researchers, instructional designers, and managers of information who collaborate as they develop educational programs with the help of their remotely located colleagues and mentors from many sectors of the community. Students will need to be able to better articulate study problems, identify and access necessary resources, and work with their peers (who may be classmates or even students in another country) to solve their problems. The roles of both the educator and learner may reverse as both explore new information resources; both will need new skills.

Even broader shifts in roles and relationships are also on the horizon. The organized involvement of parents in the education process is essen-

Box 3.3 K-12 Teacher Experiences with the Internet

- My students are beginning their ray-tracing project in computer science. Now if my school system would buy a Cray supercomputer class machine (costing approximately $20 million today), then I would have less need for telecommunications. But it is great that students in rural Wisconsin can use a supercomputer in California. This occurs because of telecommunications. Of course the argument could be made that they don't need to work in three-dimensional geometries where not only the objects can be manipulated, but also the observer. Too much like the real world.
- Our preservice teachers feel freed of many bonds that had held them previously—now they are completely free to explore the electronic libraries on the Internet, to make and maintain contacts with fellow teacher education majors at other universities, to collect and share lesson plans and ideas for teaching with other educators around the world, to ask questions of experts they are able to ferret out, to stay abreast of current events in the field, to become immensely more familiar with salient legislation, and to improve their professional literacy through communication with officers and contacts in national-level teacher organizations. A bonus is that they take great pride in the fact that they are "connected to the world." Class discussions have taken on an entirely new flavor. There is justification for more seminar-like classes in which students can have open discussions about a wide variety of topics—and this is enhanced by the remarkable quantity of information they share with each other.
- My classes have participated in a variety of educational projects on the Internet. The students have communicated with Paul Smith, a researcher from Melbourne, Australia, who was stationed at Casey Research Base in Antarctica. The students, via the Internet, were able to participate actively in Paul's research in Australia, a wonderful living lesson in science, reading, and language arts. This project will continue this fall, when Paul returns to Antarctica for 15 months of study. The children are also looking forward to returning to Antarctica, via the Internet, to share in his work.
- As a teacher, it is difficult to imagine the world of the future, yet this is the very world that I must prepare these children to enter. Change is rapid. I can only teach the children the pursuit of learning, and to develop a strong sense of inquiry, discovery, and investigation. These are the rudimentary skills these children will need to function successfully in the future. Computer telecommunications is one of the necessary tools that I need in order to prepare the children for their future. It is not a luxury.
- Regardless of how good a chemist I am, and regardless of the chemical expertise in my local area, there are always questions that I cannot answer or find answers to locally. With networking, students and teachers have an almost infinite resource pool from which to seek information and knowledge.

continued on next page

Box 3.3—*continued*

- Results that seem to be documented:

1. Holding power, especially in inner-city schools. Students get interested in computers and stay in school;
2. Reaching students with special needs;
3. Geographic awareness (through KIDLINK);
4. Foreign language skill improvement;
5. Greater sharing of ideas, lesson plans, and so on—a combination of having a sense of belonging (less isolation) and "things you can use tomorrow if not today"; and
6. Development of workplace skills.

I have rank-ordered these in terms of my perceptions, which are open to discussion.

- Virtually all of the students that I have dealt with over 15 years have been involved in chemical research. Electronic networking has provided them with electronic mentors worldwide, access to computing platforms (including supercomputers for high school students), and access to substantial on-line databases and information banks. In spite of living in a resource-rich area (Research Triangle Park, N.C.), many of my students have done projects that required resources not found in the local community.
- In working with physically impaired students, it has been clear that one of the substantial benefits of networking has been a "leveling of the playing field." Unless the student describes his or her "disability," the people he or she interacts with electronically have no concept that the person is disabled. This also applies to work with minority students or other students not from the "majority" culture.
- My all-time favorite was the use of e-mail in a junior high school "resource room." That area was normally packed with kids who were disruptive and perceived as having learning and social problems, and it was generally not a high-status area in the school. After being introduced to electronic messaging to exotic places (between Nova Scotia and the southeastern United States), the kids became really excited about writing, started attending (for the first time) to language conventions, and generally became motivated about at least one aspect of school. Here's the spin-off. Now the "low-status" resource room has a lineup of normally high achieving kids who thought the resource program suddenly had more promise. Then came the role reversal—the formerly lower achieving students became technical tutors to the formerly otherwise high achieving kids. Smiles all around.
- I have found the Internet to be the most important tool I use in my class this year. I have become very involved with Academy One, and use the various Gophers and other outside services to get information for my classes on a timely basis. For example, today I brought into class a posting of data regarding yesterday's earthquakes in Oregon. The information about the 17 larger earthquakes that had occurred in that area in 20 hours (date, time,

latitude, longitude, depth, magnitude, and notes) provided us with an excellent base of material to discuss how telecomputing can assist us on a timely basis. The story of what happened in that area . . . and how it might serve to help us prepare for our own earthquakes here in Los Angeles . . . was very clear in the data that we were able to get so quickly.

- When I first began working with fourth grade students and telecommunications, they thought that the ONLY thing an Apple II was good for was to play games like Nintendo. At the end of four months of doing on-line projects and simulations, these kids were talking about selling their Nintendos to buy "real computers." These same kids were furious to find another student from another class playing a game on the Apple II and would insist, "Get them off! I've got work to do!"
- I talked to a class about exchanging e-mail with students in other countries. At the end of class a boy came to me and asked, "Can I use Grandma's computer to write a letter to a boy in another country?" That might seem like a simple request, except that I knew the child. He lives with Grandma when his Mom isn't living with a new boyfriend. He's the kid who never turns in homework and is barely passing from year to year. The words "can I write" are like music to the ears of any teacher who wants to help students learn to read and write better. It does not matter what they are writing. If they are motivated to write something, their skills will improve as a result of the practice. That doesn't even take into account the fact that they would be learning the technology, the keyboarding skills, and information about other cultures.
- Several teachers have expressed pleasure at not needing to type materials by hand that I have made available on the network. Everything from lesson plans to laboratory exercises to curriculum development guides is now being used by classroom teachers. In most cases they can use the material quickly with a minimum of fuss on their part.
- The connection that the Internet provides is invaluable. Here I am, a research biochemist at a major university, thinking about changing to high school science teaching for the next half of my working life. Any student can access the resources of this and other labs around the world and have real-life answers to real-life questions without having to leave whatever isolated environment he or she happens to be in. We are all alive in the midst of this biological revolution, and this type of communication makes it possible for all of us to participate. Is there any real question of the educational advantage of such a system? I am no expert on the network, but I have been able to provide resources for my Earth Space Science classes that I had not even dreamed of a year ago. My students have been able to ask Danish astronomers and mathematicians questions about the history and proper pronunciation of Tyco Brahe. We have been able to get information on tropical deforestation directly from researchers in Brazil. We are now analyzing images recorded at telescopes from the summit of Mauna Kea, Big Bear Solar Obser-

continued on next page

Box 3.3—*continued*

vatory, and the Hubble Space Telescope. Yes, we might have been able to obtain the images from the Canada-France-Hawaii telescope directly without the electronic interface, but we never would have been able to make them available for student research so quickly or efficiently.

- My students have never been so excited about any technology as they are about access to the network. We are now beginning to learn together more ways to explore the net, and they look forward to obtaining their own user accounts so that they can continue to explore from their home computers. One of their first activities will be an electronic cultural exchange with students in Brazil, Florida, Central America, and Europe.
- If I participate in a workshop per month via the network, I can share most of the information to be gained by my physical attendance, and save the taxpayers over $10,000 in air fares at the same time.
- Our sixth graders needed current information about countries on which we have some historical materials but few that are up to date. So we went to the CIA World Factbook (Outside services, TENET> open 137.113.10.35—William and Lee; login: lawlib; search CIA World Factbook; of country (Egypt, India, Israel, West Bank, Gaza Strip, Greece, Italy, Syria, etc.)). The students needed information about what was formerly Mesopotamia, plus the other countries. The available information about each included area, population, government, transportation, economics, general information, languages, religions, money, and more. The Factbook was the 1992 edition, and it gave better information than we could have obtained from hundreds of dollars worth of books about these countries. We were searching immediately after the meeting between Israel and the Palestinians, and so the information was especially pertinent. The students were so excited to find so much information on every country they searched for. To test the system, they went on to look up other countries not on their list—most of which were part of their own heritage.
- This message is to let you know that as an elementary teacher I can no longer survive without e-mail! E-mailing is a truly wonderful learning tool for teachers and students. Students who have never had a desire to read much or write anything are now very interested in doing both because of e-mail. They see how important it is to be able to read so that you can read messages concerning your project, directions for the computer, and sources of information for your project both on and off the computer. They realize how important it is to clearly communicate your data on a project. Even punctuation and capitalization have become important to them.

NOTE: These anecdotes were compiled from electronic mail messages sent by educators in response to an electronic request by a member of the study committee. Educators who use the Internet to varying degrees and for diverse purposes were asked to document how network use has affected teaching and learning.

tial and will occur more frequently and naturally through networks that connect the home to the school and to libraries and other repositories of information. In time, the teacher-parent relationship will change as teachers take on new roles as facilitators who broker research and monitor student progress, often assisted by specialists, such as nutritionists, who provide guidance in areas that affect student behavior and performance. Another contributor, made possible by the Open Data Network, will be the employer of the future. The needs of the workplace, rather than mere job listings, will play an integral role in defining the curricula for a select number of students, especially those seeking vocational or professional pursuits after college.

Realizing process-change benefits will not be automatic. Compounding insufficient physical resources have been cultural barriers. For example, schools tend to be managed as a hierarchy with implicit (and explicit) expectations about who gets information first; collaboration with colleagues, especially those outside one's school building, is not the norm. Open, especially student-driven, communications challenge those expectations and (in the behavioral sense) protocols. Finally, telecomputing projects have been seen as an add-on, requiring extra effort. Changes in schooling require support and involvement from all segments of the education hierarchy—the administrators, parents, community, support staff, and classroom practitioners. Provision of the broad NII envisioned in this report is an important step in this direction. *But until a systemic approach is taken that applies to all types of education professionals and activities and addresses the requirements of implementing a new communications system, much work will be needed to truly integrate networking and information infrastructure into schooling.* Accordingly, leading education professionals, their organizations, and some policymakers have begun to explore the linkages between educational reform and the emerging NII.[28]

The absence of consensus on how to incorporate network-based applications into education suggests a need for research into appropriate technology, appropriate levels for providing networking and internetworking support, the economics of school networking (funding, policies), content standardization, and information-age skills and knowledge assessments. As one workshop report observed, "Educators will not use networks to teach just because the technology is pervasive in society."[29]

Initial federal programs have concentrated on computer and networking activities relating to mathematics and science; for example, that is the focus of the NSF activities through the Education and Human Resources Directorate, including the National Infrastructure for Education (NIE) program (which also involves NSF's Computer and Information Science and Engineering Directorate). Yet "educating the whole child" or the process of education as a whole extends beyond math and science; the process of communicating over a network has been shown to have benefits in the areas of reading and writing, for example (see Box 3.3). Also, true NII support for education must encompass administrator, principal, and teacher preparation; public awareness; and other facets of the education process that may not be captured in programs focusing on enhancing science and math education.

Several states have taken a leadership role by potentially eliminating restrictive state provisions in order to facilitate and support the development of teacher and administrator preparation programs and to assist in the restructuring process. In Texas this effort integrates technologies and innovative teaching practices into preservice and staff development training of teachers and administrators. Through regional Centers for Professional Development and Technology—a collaboration among universities, school districts, regional education service centers, and the private sector—systemic change is occurring in the professional development programs.

The local basis for educational funding contributes to substantial differences in the ability of schools and school districts to make use of networks today. Several states have moved to augment local capabilities; for example, the state of Florida has implemented a grants program to retrofit schools. In Texas the state legislature implemented the Technology Allotment Fund, which provides an allocation that districts may apply for if they submit a plan for the integration of technology. The amount is $30 per child and is based on the average daily attendance of the school district. There are also equity concerns among individual students, only some of whom may be able to afford their own, home-based technology. Where access has been achieved, teachers report that disadvantaged students have gained access to resources and have responded far beyond what has been observed with more conventional educational approaches (see Box 3.3).

Equally important will be the outreach capability of the networks to ensure that the poor or rural residents of a community are not excluded.[30] Realizing that potential, however, is a function of time: the NTIA Telecommunications and Information Infrastructure Assistance Program can make only one contribution, the NSF Connections program and NIE program two others, and it may be several years before all K-12 schools (of which there are more than 110,000) are connected to the Internet. As in the case of higher education, a variety of organizations have emerged to pursue opportunities relating to K-12 networking. These include the International Society for Technology in Education (ISTE) and the Consortium for School Networking (CoSN).

LIFELONG EDUCATION

There is consensus across the research and industrial communities that people should expect to be retrained periodically throughout their careers. An NII is an attractive vehicle for this purpose.

In some career fields the relevant base of information is doubling every three or four years, a situation that is redefining what is meant by "certified" or "qualified" as regards professional competency. This is perhaps most apparent in areas that are affected by science and engineering or are heavily positioned within the regions of public interest and subject to regulatory controls. An example is medicine and health care, where change is continuous and ranges from new, or upgraded, government regulations to the latest developments in bioengineering. Given this avalanche of information, "professionalism" may some day be defined in terms of a trained user's access to and familiarity with "search" procedures that are friendly and analytical, and that enable the user to acquire information efficiently on a need-it-now basis or to perform serendipitous browsing.

Self-education, beyond formal school educational programs, will be accomplished by individuals at ease with networking procedures that permit easy access to on-line digital libraries and make available virtual trips to museums and science projects without leaving home. Continuing education via teleconferencing will enable the learner to participate in preprogrammed or live academic courses remotely and to receive academic credits, if desired. Acceptance of the latter by educational institutes may, however, bring to the surface a wide range of issues regarding accreditation and the changing role of the institution and the instructor in the education process.

The NII will create important changes in professional-layperson relationships. As nonprofessionals gain increased access to databases on the network, they will begin to increase their awareness of a multitude of topics that previously were the strict provinces of the professional. The long-term result will be the creation of a much better informed citizenry. The potential for information infrastructure to support paraprofessionals and allied health professionals in medicine, for example, is already in evidence. An extension of this facility will be seen in the growing number of information entrepreneurs providing information for a fee to laypersons on selected professional subjects using such resources as electronic bulletin boards and specially designed electronic libraries.

LIBRARIES AND THE BROADENING OF PUBLIC INTEREST NETWORKING

Libraries complement both research and education. They figured first into NREN and now into the vision of an integrated, broadly useful

Box 3.4 Contributions of Public Libraries

The United States has the world's most extensive public library system, with some 15,482 physical locations, including branches. In the past year, more than half (53 percent) of the adult population used a public library, as did 74 percent of children aged 3 to 8. A picture of the role and impact of public libraries can be derived from statistics presented in "America's Libraries: New Views in the '90s," a four-page 1993 update to *America's Libraries: New Views*, a special report published by the American Library Association, Chicago (1988).

Users of public libraries are diverse. A 1991 household survey by the National Center for Education Statistics revealed that 42 percent of African Americans said they had used a public library in the last year, as did 32 percent of Hispanics, 55 percent of whites, and 52 percent of all others. Expenditures for collections and services total $4.3 billion annually, or about $17.80 per person, and represent less than l percent of all tax dollars. Support for public libraries comes primarily from local communities. In 1991, sources of funding were approximately 76.8 percent local, 13 percent state, 1.2 percent federal, and 9 percent other sources.

The information resources available to and used by the public are vast. More than 1.4 billion books, magazines, video and audio tapes, computer software, and other items are borrowed each year. Public library circulation increased 5 percent in 1991, the latest in a steady series of increases over the last decade. Some 222 million reference questions were answered by public librarians in 1991.

The general public itself has already identified the library as a key provider of information. Participants in a 1992 Gallup poll indicated the following as "very important" roles for public libraries:

- Formal education support center (90%),
- Independent learning center (83%),
- The preschooler's door to learning (82%),
- Research center (67%),
- Community information center (63%),
- Reference library to community businesses (54%),
- Popular materials library (50%),
- A comfortable place to read, think, or work (49%),
- Reference library for community residents (47%), and
- Community activities center (40%).

The number of public libraries offering electronic information services is growing rapidly; according to a survey conducted by Opinion Technology, more than 80 percent of all public libraries and 99 percent of all academic libraries now offer such services. Computer-related services are available to the general public in the following percentages of public libraries serving populations of 100,000 or more:

Box 3.4—*continued*

- CD-ROM databases (79%),
- Remote database searching (71%),
- Microcomputers (62%),
- Microcomputer software (57%),
- On-line public access catalog (60%), and
- Dial-up access to on-line catalog (29%).

NII. The High-Performance Computing Act of 1991 (PL 102-194) envisioned libraries as both access points for users to utilize the network as well as providers of information resources via the Internet.[31] The administration's characterization of the NII carries forward this expectation, expanding it to include a training function.

Libraries have long played a central role in society as providers of information resources (hence the dream of digital libraries[32]), as points of access to information (information that is increasingly in electronic form), and as interfaces with the end user (librarians teach users how to locate and interpret information) for many constituencies. The United States has about 125,000 libraries, including public, academic, research, and special libraries (Box 3.4). Research and academic libraries resemble the research and higher-education communities that they serve in terms of problems and prospects; public libraries resemble K-12 education in terms of severe financing conditions and service to a broad community.[33] Also, as in K-12 education, public libraries often are limited by inadequate physical infrastructure and training, and they tend to be extremely sensitive to cost.[34]

In the NREN environment, research and academic libraries have taken a leadership role in advancing network-based initiatives to provide access to information resources in support specifically of research and higher education through interconnection and interlibrary loans. They also support wider segments of the public.[35] Reduced operating hours at local public library facilities, reflecting budget constraints, have shifted demand to research and academic libraries, a shift enabled in part by network connections. Cooperation among different types of libraries extends beyond resource sharing to such activities as preservation, education, and training, especially relating to network-based applications. For example, the "CIC" libraries (components of several primarily midwestern state universities) have been coordinating on plans for acquisition, storage, cataloging, preservation, and retrieval relating to over 600 electronic journals, the contents of which will be supplied to member institutions over the mid-level CICnet.[36]

The nature and mix of services provided by libraries are changing with the evolution of information infrastructure. Recent Association of Research Libraries' statistics indicate that research libraries are moving from the "just-in-case" model of on-site resources to the "just-in-time" model of resource-sharing.[37] As physical acquisitions costs for scholarly information in print form mount for libraries, interest in "no-fee" (or low-cost) information through the Internet grows as well. Printed service subscriptions and monograph acquisitions are declining.[38] Internet access enables libraries to leverage their resources so as to acquire more of the scholarly record, and to own materials collectively and share them between libraries and their end users (Box 3.5). On the other hand, it is not without its difficulties; Box 3.6 outlines barriers to broader library activities involving the Internet.

In the years ahead, libraries will communicate and provide access to information in a variety of formats—digital, voice, graphic—and employ multimedia technologies via a ubiquitous and seamless web of interrelated networks. Public access programs and policies proposed and implemented today will be central to this emerging information infrastructure.

Libraries participate in numerous experiments and pilot programs that demonstrate the utility of high-capacity networks for the exchange and use of information for all disciplines (Box 3.7). Research and academic libraries already engage in and/or provide electronic document delivery, electronic journals, full-text databases, end-user searching, training, network access, development of network navigational tools, Online Public Access Catalog enhancements, cooperative development of databases and hardware, and policies, services, and strategies that promote access to information in lieu of ownership. There are numerous discussion databases and electronic forums developed by and aimed at library professionals, who access them through the Internet. The tradition of service to those who cannot afford books and other sources of information is being and can be further expanded in libraries to include not only local access to terminals and other network-access devices but also possibly even loan of such devices.

Network applications in libraries today focus on access to resources such as books, journals, and on-line files; in the near future, the focus will be on access to and use of research materials and collections generally inaccessible but of extreme research value, including photographs, satellite and related spatial data, archival data, videos and movies, sound recordings, slides of paintings and other artifacts, and more. This development presumes that digital facsimiles or digital records are available for access and distribution over a network.[39]

The development of such digitized resources is central to the concept of digital libraries. Characteristics of digital libraries include the following:

Box 3.5 Advantages of Internet Access via Public Libraries

• *Equitable and ubiquitous access provider.* Libraries offer access to the network; the equipment, technical support, and software needed to access it; and the information resources available through it.

• *Affordable access.* Library experience with current commercial electronic information services indicates that several services tend to be priced for the institutional or corporate subscriber, and not for the individual user who may have only occasional need for access to a small portion of an information source or database. In this environment, the public library provides the electronic equivalent of one of its traditional functions—to provide access to a wide variety of information sources and viewpoints regardless of a user's economic status or information-seeking skills.

• *Network information resource provider.* A 1992 journal article* identified some of the databases developed by and uniquely available from public libraries: community-based information and referral files listing government and social services; query files of questions frequently asked by the public, with answers; genealogy files for specific local geographic areas; local newspaper indexes; annotated reading lists; catalogs of holdings; tour and day-trip itineraries for local historical sites; and so on.

• *Access to government information.* Libraries have long had responsibilities under law and custom for partnering with governments to provide public information. This has been accomplished through the federal Depository Library system, with libraries in every congressional district, and state depository systems. The library as the local access point for electronic government information is a natural extension of this partnership.

• *Training and assistance for the public.* Unlike most sites for public access terminals (which range from government buildings to universities, from shopping malls to laundromats), public libraries have trained staff available for consultation and training in the use of the library's resources, including electronic information resources. A logical extension is to provide training for the public in the use of networks and networked information resources plus point-of-use consultation, guidance, and technical assistance, as well as to develop on-line training and interpretative aids.

• *Library as electronic gateway.* Libraries of all types have more than 25 years of experience in using computer and communications technologies to link together to share bibliographic information for cataloging and interlibrary loan. The Internet and the National Information Infrastructure have the potential to link libraries further for electronic sharing of full-text, graphic, and multimedia library resources; to link library personnel electronically for new kinds of reference services; to link libraries to nonlibrary sources of information; and to provide access to the local library from any location with a computer and modem.**

continued on next page

Box 3.5—continued

---

*Isenstein, Laura J. 1992. "Public Libraries and National Electronic Networks: The Time to Act Is Now!" *Electronic Networking* 2(2, Summer):2-5.

**An excellent discussion of the advantages of Internet access via public libraries—and the source from which this box was derived—can be found in: Henderson, Carol C. 1993. "The Role of Public Libraries in Providing Public Access to the Internet," prepared for a conference, Public Access to the Internet, John F. Kennedy School of Government, Harvard University, May 26-27, 1993.

- *Large size:* The total of all printed knowledge is doubling every eight years, and many research databases dwarf past collections of information;
- *Manipulability:* The use of an electronic digital format means that data of any kind can be potentially communicated, analyzed, manipulated, and copied with ease;
- *Inclusion of mixed media:* The digital library will consist of multiple forms and formats of information including images, sounds, texts, computer programs, and quantitative data;
- *Distributed:* The digital library is not a single entity or database in a specific geographic location. Instead it consists of resources that are constantly changing and available on a distributed basis. The evolution of the digital library and its distributed nature are fundamental characteristics relating to the digital library's value to the user; and
- *Accessibility and interactivity:* Digital libraries will be accessible to new communities and a wider-range of users. The resulting availability of new research and new knowledge will in turn increase the value of the digital library, a benefit that will come from the interactivity between the user and the digital library.[40]

Creation of digital libraries will likely exacerbate information management and policy issues and will require additional research to resolve problems that may thwart progress (see Chapter 4).[41] Many if not all of the information policy issues requiring attention are not new. They relate to freedom of expression, intellectual property, access, privacy and confidentiality, security concerns that include the integrity and reliability of the date resources, and the preservation and archiving of data resources.[42] The nature of the technologies either exacerbates existing tensions (e.g., relating to copyright and fair use) or presents new questions and opportunities to rethink existing practices.

Notable among the technical issues relating to library participation in an NII is the need for librarians—working together with publishers and

Box 3.6 Barriers Facing Public Libraries as Public Access Points

- *Few public libraries on the Internet.* The first major barrier to public libraries as public access points on the Internet is that so few are connected currently, perhaps on the order of a few hundred out of 9,000 (located at 15,000 sites, including branches). Gaining access has not been easy. University computer centers closely monitor their "guest accounts." The primary mode of access offered by most of the regional affiliates and networks, as well as private providers, requires a direct high-speed line, with the user serving as a full node on the network. For most libraries, even in relatively large municipalities or corporations, the finances and other resources required for start-up of dedicated line operation are simply too great. While there are some less expensive dial-up options available, most of these offer electronic mail only and do not provide FTP and the telnet functionality that is especially important to libraries. Increasingly, library or library-related networks—cooperative, not-for-profit regional or state-based library service organizations that broker Online Computer Library Center electronic bibliographic network services and/or other technological services for groups of individual libraries—are becoming the providers or brokers of Internet connectivity.*
- *Lack of affordable access.* Dial-up, entry-level connectivity is only the beginning of the cost for libraries, many of which will also incur long-distance charges. A good example is found in the state of Wisconsin, where it costs about $14,000 annually to be a full member of WiscNet, the state network providing Internet access. For an entire campus, that is not a major expense. But 45 percent of U.S. public libraries have budgets under $50,000; in Wisconsin it is 54 percent. For such libraries, $14,000 annually is a major expense. In addition, these costs probably do not include local staffing costs for technical support, user support, and training.
- *Lack of user-friendly interfaces.* Lack of user-friendly interfaces and tools, the lack of databases and resources geared to public use, and the seemingly limitless information on the Internet, not all of it useful to public libraries and their users, constitute additional barriers. The lack of organization of much of the information on the network is another serious barrier to using it effectively.
- *Training and support needed for staff and the public.* Librarians need to know what is available, how to find it, and what the technological problems may be and how to solve them, before using the Internet on behalf of users or providing direct access to the public. Persuasion is needed that a new set of procedures and costs will be worth the investment of scarce time or money. Some of upper library management is not conversant with the latest technologies. In other cases, librarians need ammunition to convince parents, school boards, or local governments that are themselves technologically out of date or challenged. Especially important is the need to train staff before offering Internet service for public access.

continued on next page

Box 3.6—*continued*

• *Policy issues.* As public libraries begin to move toward providing direct public access to the Internet, they face a host of administrative and policy issues that must be addressed. In general, these are not unique to public libraries, and so they are listed but not discussed. They include:

- Privacy and security issues,
- Intellectual property protection and fair use of copyrighted materials,
- Affordability of commercial information services,
- Interoperability and standards,
- Whether and how to impose limits on public use when demand outstrips resources (time, equipment, capacity, or budget),
- Scalability of Internet address system, and
- Censorship, access by minors.

Tools and policies developed by the library field and put to regular use in libraries can help in coming to grips with these various problems.

---

*An excellent discussion of the barriers facing public libraries—and the source from which this box was derived—can be found in: Henderson, Carol C. 1993. "The Role of Public Libraries in Providing Public Access to the Internet," prepared for a conference, Public Access to the Internet, John F. Kennedy School of Government, Harvard University, May 26-27, 1993.

other information providers, research users, and information and computer scientists—to develop standards, common formats, and controls that will permit users to identify, locate, and access needed resources in a consistent fashion. Standards and protocols will also be needed for those users of digital libraries who may lack the needed skills to effectively utilize the networked environment and who may not have a librarian to call upon.

Broad use of digitized resources outside of a library facility with professional staff is expected to increase with home-based and other remote access to information infrastructure. Overall, the future role of libraries will evolve to reflect and interact with developments in individual, personal information retrieval systems and also developments in the publishing arena and other sources of supply for electronic information resources.

Box 3.7 Experimental Library Projects

• *The Electronic Text Center and On-Line Archive of Electronic Texts at the Alderman Library, University of Virginia.* Opened in 1992, the Electronic Text Center combines an on-line text archive with computer hardware and software systems for the creation and analysis of text. The archive, including electronic texts encoded with Standard Generalized Markup Language, includes the entire corpus of Old English writings, several hundred Middle English and Modern English works, and smaller selections of French, Latin, and German works. Considerable effort has gone into creation of on-line and printed documentation, much of it available over the Internet via gopher and World-Wide Web servers. Over 7,500 remote logins from over 1,600 on-line users were counted in 1993. The electronic texts have been used in a wide range of research and educational activities.

• *The North Carolina State University Digitized Document Transmission Project.* Scanned images are transmitted over the Internet to libraries, researchers' workstations, and agricultural extension offices. Collaborating on this project are the National Agricultural Library and 14 land-grant university libraries in 11 states.

• *The Economic Development Information Network, or EDIN.* This collaborative effort between Pennsylvania State University, the Pennsylvania State Data Center, and the Institute for State and Regional Affairs provides access to bulletins and news releases, issues of *Commerce Business Daily*, directories of economic development centers and agencies, database files pertaining to demographic and economic data, and more.

• *The Chemistry Online Retrieval Experiment (CORE)*, a prototype electronic library of 20 American Chemical Society (ACS) journals that is disseminated over Cornell University's local area network. The project is a collaboration among four participants—the ACS and its Chemical Abstracts Service division; Bell Communications Research (Bellcore, the research arm of the regional telephone holding companies), Morristown, N.J.; Cornell University's Mann Library; and the Online Computer Library Center. The CORE system enables Cornell faculty and students to search a database that eventually will include more than 10 years' worth of issues of 20 chemical journals and information from scientific reference texts. Users can retrieve articles electronically, complete with illustrations, tables, mathematical formulas, and chemical structures. They also can switch to articles on related topics, or to referenced articles, using hypertext-type links. The database is constructed with the same composition data used to publish the print-on-paper versions of the journals, thus minimizing the labor needed to create it and keep it current.

• *TULIP (The University Licensing Program).* This project of Elsevier Science Publishers is a database of 42 Elsevier-published science journals. Researchers at 17 participating universities, including the nine campuses of the University of California system, can access these journals over local area networks and print journal pages or articles on demand. The system provides access to bitmapped page images (for viewing), full-text files (for searching), and bibliographic files.

CROSS-CUTTING OBSERVATIONS

The key parameter for the research, education, and library communities is cost. Limited ability to pay for services is typical; a related factor is a need for predictable charges because of the limited, fixed budgets of schools and research institutions. Higher costs may mean a reduction, elimination, or preclusion of access for those actual and potential users with the least robust funding—K-12 education, smaller and less affluent institutions, smaller and more poorly funded departments or researchers—absent targeted support. Higher—or in some cases any—costs will force users and their sources of funding to confront the issue of whether there is a trade-off between networking and their "basic" activity (research, education) or whether networking is so connected to their basic activity that they would prefer to reduce spending elsewhere to support it.

At the time the NREN concept was conceived, some members of the research community characterized funding as zero-sum: for those with limited or no use of networks, money for networking was perceived as money lost to research.[43] Although the benefits of networking are now more broadly appreciated in the research community, the transition from a "free" service to a fee for service revives questions about possible trade-offs and how integrated networking is or should be in research or education. The problem may be particularly acute in K-12 education, where the institutional barriers to commercial fulfillment of a public service must be addressed by public institutions.

Higher costs should also prompt some consideration of possible gains in efficiency. Efficiency does not receive the kind of attention in the research community, in particular, that it receives in industry. Information access is one way in which networks increase efficiency in education, although such a classical benefit is not the principal benefit in an area where the dominant cost is an already leveraged resource: personnel and teachers.[44] On the other hand, this situation underscores the need for training and skill development, which will affect how much benefit is received, and how fast.

In research and education, outputs, inputs, and the relationship between them are hard to characterize and control. Yet cost savings can be an important benefit of the use of information infrastructure, because they are inherent in the notion of networks and information infrastructure as shared resources. That sharing enables broader use of resources than would be possible if each researcher, educator, librarian, or student had to be individually capitalized. In addition, as the Internet experience demonstrates, information infrastructure can facilitate cross-sectoral sharing, including the building of bridges between K-12 education and higher education and research. For example, many states are now working to connect universities, colleges, community colleges, K-12 schools, and public libraries via networks to support both learning and research.

It would be a mistake, however, to frame NII planning for the research, education, and library communities simply from the perspective of financial aid requirements. These communities have been information providers as well as consumers, and they will continue to make important contributions to network information resources in this regard in the future. The fact that these communities typically do not charge for their information services might be considered an important factor in considering how to charge them for their network access. Another important factor may be the fact that these communities also actively train their constituents in network use.

Viewed from a national and even a global perspective, research and education in all fields will continue to provide the bedrock for U.S. social and economic growth. Competitiveness in international markets, education of all members of society for positive and meaningful participation in the coming decades' tasks and opportunities, and breakthroughs in scientific and nontechnical scholarly pursuits that will improve our way of life are among the goals that can be fostered by ensuring equitable access to the NII. The steps we take now to realize the potential benefits of the information future will be a significant measure of the country's progress in shaping its agenda to reflect its ideals.

NOTES

1. In networked environments, museums, archives, and other information providers are expected to play an increasing role, complementing libraries.

2. CSTB has planned, together with the Institute of Medicine, a comprehensive study of information infrastructure for health care.

3. Peters, Paul Evan (Executive Director of the Coalition for Networked Information, Washington, D.C.). "Responses to Questions," message (electronic mail system) to Susan L. Nutter, August 13, 1993. Stone-Martin, Martha, and Laura Breeden. 1994. *51 Reasons: How We Use the Internet and What It Says About the Information Superhighway.* FARNET Inc., Lexington, Mass.

4. For example, the electronic physics publishing activities centered at the Los Alamos National Laboratory are recognized as helping to democratize the field, allowing new insights to go beyond the "old boy network" that previously was the only group to know of key results when they were still new.

5. TOGA-COARE refers to the decade-long Tropical Ocean-Global Atmosphere international research program, including the Coupled Ocean-Atmosphere Response Experiment.

6. Pool, R. 1993. "Networking the Worm," *Science* 261:842.

7. Computer Science and Telecommunications Board (CSTB), National Research Council. 1993. *National Collaboratories: Applying Information Technology for Scientific Research.* National Academy Press, Washington D.C.

8. Cuticchia, A.J., M.A. Chipperfield, C.J. Porter, W. Kearns, and P.L. Pearson. 1993. Managing All Those Bytes: The Human Genome Project," *Science* 262:47-48.

9. Carey, John. 1994. "Could America Afford the Transistor Today?" *Business Week*, March 7, pp. 80-84.

10. CSTB, 1993, *National Collaboratories*.

11. CSTB, 1993, *National Collaboratories*.

12. The Branscomb report recommended broadening access to scientific and engineering workstations for NSF's 20,000 investigators and contemplated increasing simulation and visualization activity using personal computers and more powerful systems. These desktop systems, in turn, require high-performance input-output, distributed access to databases, and other infrastructure-related complements. Branscomb, Lewis, et al. 1993. *From Desktop to Teraflop: Exploiting the U.S. Lead in High Performance Computing*, NSF Blue Ribbon Panel on High Performance Computing. National Science Foundation, Washington, D.C., August.

13. One concern already being raised is whether those with greatest proximity to a vBNS node will have easier access.

14. "Scholars quickly understand that electronic documents have several obvious benefits: they can be searched quickly for phrases, words, and combinations of words, allowing one to try out notions and hypotheses with great speed; they encourage large-scale searches over oeuvres, genres, and centuries, searches that are difficult and time-consuming with printed texts alone; they can provide access to texts otherwise unavailable, and they allow such work to be done from one's home or office." Seaman, David. 1993. "Gate-keeping a Garden of Etext Delights: Electronic Texts and the Humanities at the University of Virginia Library." *Gateways, Gatekeepers, and Roles in the Information Omniverse: Proceedings from the Third Symposium*, November 13-15, 1993, Washington, D.C.

15. Tibbs, Helen R. 1991. "Information Systems, Services, and Technology for the Humanities," *Annual Review of Information Science and Technology* (ARIST) 26:287-346.

16. Peters, Paul Evan, "National Initiative on Arts and Humanities Computing," message (electronic mail system) to CNI-Announce subscribers, January 1, 1994.

17. Mandelbaum, Richard, and Paulette A. Mandelbaum. 1992. "The Strategic Future of the Mid-level Networks." Pp. 59-118 in *Building Information Infrastructure*. Harvard University Press, Cambridge, Mass.

18. Branscomb et al., 1993, *From Desktop to Teraflop*.

19. In the early 1980s Carnegie Mellon University and the Massachusetts Institute of Technology were the pioneers in incorporating technologies that wove telecommunications into the instructional process. Oklahoma State University now uses telecommunications to teach German to students in rural high schools. Chico State University provides a range of courses to sites in rural northeastern California as well as a graduate program in computer science to industry locations across the nation. State systems in Oregon, Utah, Texas, and California are planning major telecommunications efforts as a means to increase student access to and utilization of resources. An advisory commission to the California community colleges recently recommended the expansion of telecourses to increase the productivity of the system.

20. This initiative was discussed by federal agency personnel in 1988, and it was urged by EDUCOM, and its Networking and Telecommunications Task Force, in 1989; it had antecedents in proposals to the National Science Foundation by computer scientists Robert Kahn and Vinton Cerf. Mandelbaum and Mandelbaum, 1992, "The Strategic Future of the Mid-level Networks"; personal communication, Stephen Wolff, National Science Foundation, June 1993; Lynch, Clifford A., and Preston, Cecilia M. 1990. "Internet Access to Information Resources," *Annual Review of Information Science and Technology* (ARIST) 25:296.

21. The Fielding Institute represents a higher-education institution that is built around

the virtual campus, serving a geographically distributed student body dominated by adult professionals pursuing graduate and doctoral programs.

22. Burr, Elizabeth. 1994. "Electronic Studios," *News from FONDREN* 3(3, Winter):1-3.

23. A Macintosh-based, real-time, multiparty videoconferencing program, CU-SeeMe, is available free from Cornell University under the copyright of Cornell and its collaborators. CU-SeeMe version 0.60, with an improved user interface, provides a one-to-one connection or, by use of a reflector, a one-to-many, a several-to-several, or a several-to-many conference depending on user needs and hardware capabilities. It displays 4-bit gray-scale video windows at 160 × 120 pixels or at double that diameter, and does not (yet) include audio. At this time CU-SeeMe runs only on the Macintosh using an IP network connection over the Internet. A PC version is under development and is expected soon. With CU-SeeMe each participant can decide to be a sender, a receiver, or both. Personal communication, Jill Charboneau, Cornell University, April 1994.

24. Also connected are 10 K-12 schools or school districts, 14 state or federal government agencies, 11 health care organizations, 34 nonprofit organizations, and 13 businesses. Personal communication, Eric Aupperle, Merit Inc., April 13, 1994.

25. Personal communication, Eric Aupperle, Merit Inc., April 13, 1994.

26. Klingenstein, Kenneth. 1993. "The Boulder Valley Internet Project: Early Lessons in Early Education," *INET '93 Proceedings*, pp. ECA 1-8.

27. Klingenstein, 1993, "The Boulder Valley Internet Project."

28. For example, the National Coordinating Committee on Technology in Education and Training (NCC-TET), which includes a large number of K-12 and higher-education organizations as well as industry and government members, recently issued a statement on educational needs associated with the NII. It noted that "the NII (as it develops) and related technologies can be key supports for education reform and therefore be incorporated into education reform initiatives at the national, state, and local levels." National Coordinating Committee on Technology in Education and Training. 1994. "The National Information Infrastructure: Requirements for Education and Training," March 25, electronic communication.

29. "The CoSN/FARNET Project, Building Consensus/Building Models for K-12 Networking." n.d. Report of October 28, 1993, workshop, electronic distribution.

30. Telecomputing can enhance interactions within communities in the context of strengthening K-12 education. In Texas, for example, where 70 percent of the counties are medically underserved, the means to link existing infrastructure and higher-education institutions with schools to share information about health, education, and medical care is provided by a state network developed for education, TENET. One collaborative initiative links TENET and the South Texas Center for Preventive Genetics at the University of Texas. A pilot project has begun to develop a monitoring and treatment system for children diagnosed with certain birth defects. An integral part of the project is communication with school nurses, school clinics, and teachers via TENET. For example, teachers and nurses are asked to become involved in monitoring compliance with dietary requirements at school. The goal is to remove distance as a barrier to effective treatment, improve prognoses for individuals being undertreated, increase dietary treatment compliance, and provide a database to answer researchers' queries about efficacy of treatment as it relates to cognitive function.

31. "The Network is to provide users with appropriate access to high-performance computing systems, electronic information resources, other research facilities, and libraries. The Network shall provide access, to the extent practicable, to electronic information resources maintained by libraries, research facilities, publishers, and affiliated organizations." PL 102-194, section 102.

32. Note that in an environment filled with jargon, it is important to distinguish between

a virtual library—a facility for accessing information, which can be stored in various forms or media—and a digital library, which usually refers to a collection that is itself digitized. The virtual library is more easily implemented.

33. "Public libraries" means public libraries as defined in the Library Services and Construction Act, state library agencies, and the libraries, library-related entities, cooperatives, and consortia through which library services are delivered.

34. McClure, Charles R., Joe Ryan, Diana Lauterback, and William E. Moen. 1992. *Public Libraries and the INTERNET/NREN: New Challenges, New Opportunities.* School of Information Studies, Syracuse University, Syracuse, New York.

35. According to Clifford Lynch,

> Work in [the interlibrary loan] area has ranged from the use of electronic ILL systems linked to large national databases of holdings (such as OCLC) which allow a requesting library to quickly identify other libraries that probably hold materials and dispatch loan requests to them through the exploitation of technology to reduce the cost of the actual shipment of material. The first step in this latter area was for the lending institution to send a Xerox of a journal article rather than the actual journal copy, so that the borrowing library did not have to return the material and the lending library did not lose use of it while it was out on interlibrary loan. . . . More recently, libraries have employed both fax and Internet-based transmission systems such as the RLG Ariel product to further speed up the transfer of copies of material from one library to another in the ILL context, and with each additional application of technology the publishers have become more uncomfortable, and more resistant (with some legal grounds for doing so, though again this has not been subject to test). Interestingly, over the past two years, we have seen the deployment of a number of commercial document delivery services (the fees from which cover not only the delivery of the document to the requesting library but also copyright fees to the publisher) offering rates that are competitive—indeed, perhaps substantially better—than the costs that a borrowing library would incur for obtaining material such as journal articles through traditional interlibrary loan processes. . . . Now, consider a library acquiring information in an electronic format. Such information is almost never, today, *sold* to a library (under the doctrine of first sale); rather, it is *licensed* to the library that acquires it, with the terms under which the acquiring library can utilize the information defined by a contract typically far more restrictive than copyright law. The licensing contract typically includes statements that define the user community permitted to utilize the electronic information as well as terms that define the specific uses that this user community may make of the licensed electronic information.

See Lynch, Clifford A. 1993. *Accessibility and Integrity of Networked Information Collections.* Background Report/Contractor Report prepared for the Office of Technology Assessment, July 5, p. 16.

36. Shaughnessy, Thomas W. 1994. "Libraries Organize as a Virtual Electronic Library," *Library Line* 5(March):1-2.

37. Association of Research Libraries. 1992. "Key Issues to Consider in NREN Policy Formulation," *Proceedings of the NREN Workshop,* Monterey, Calif., September 16-18. Interuniversity Communications Council Inc., Washington, D.C.

38. The 100 members of the Association of Research Libraries paid $19.8 million more for services in 1992-1993 than in the previous year but purchased 30,000 fewer subscriptions and 30,000 fewer monographs. Association of Research Libraries (ARL). 1994. "State of Research Libraries and Importance of HEA Title II-C to Research and Education." ARL, Washington, D.C., March 2.

39. These records and network-based access to them present both technical and intellectual property management issues. See Library of Congress. n.d. "Delivering Electronic Information in a Knowledge-based Democracy," Summary of Conference Proceedings, July 14, 1993.

40. These characteristics of digital libraries were presented in a statement by the Associ-

ation of Research Libraries to the Subcommittee on Science, Committee on Science, Space and Technology, from the Hearing Record of February 2, 1993, regarding the High-Performance Computing Act of 1991.

41. A new NSF-ARPA-NASA research program associated with the NREN program will fund relevant research.

42. Recent Supreme Court decisions relating to access and the availability of government electronic records underscore the need for rules and regulations that govern their preservation and disposition. Building concerns about archiving and preservation into existing network projects will be important.

43. See Computer Science and Technology Board, National Research Council. 1988. *Toward a National Research Network*. National Academy Press, Washington, D.C. (The Computer Science and Technology Board became the Computer Science and Telecommunications Board in 1990.)

44. Note that libraries may have greater integration of networking for public access and internal operational purposes than is typical of K-12 education.

4

Principles and Practice

Discussions of a national information infrastructure (NII) and discussions of research and education networking tend to embrace several tenets or principles for the design and implementation of information infrastructure.[1] There are different classes of principles: some relate to the availability of information infrastructure (e.g., ubiquitous and equitable access); some relate to the terms and conditions of use (e.g., protections for privacy and security). In some instances, desired attributes of the structure of the infrastructure are also presented as principles. In this report, those technical issues, including security protections, are discussed separately as part of a comprehensive consideration of open information networking (Chapter 2).

In the past year, the administration, industry, and professional and other advocacy groups have presented sets of principles intended to guide the development of the NII (see Appendix B for samples of such principle sets). These principles reflect value judgments and suggest the directions in which many individuals and groups will try to direct the development of architecture and, in particular, the deployment (including investment, pricing, and financing) of the NII.

On the surface there is significant commonality among the lists that have begun to circulate, reflecting the universal quality of several principles, the prominence of public interest advocates in their formulation, and the fact that it is relatively easy to achieve agreement on high-level statements. For example, the presence of privacy, security, and intellectual property protection across various lists of principles shows that these issues not only reflect shared mores and norms, but also are seen as broad enablers of information infrastructure. At the same time, there also are differences among these lists in terms of emphasis as well as content. Further, the meaning or interpretation of these principles may differ for different groups.

The research, education, and library communities have valuable experience with the implementation, or lack of it, of various principles in the Internet context. Given the nature of their missions and reliance on financial support, their perspectives and concerns will differ from those of commercial or consumer users of information infrastructure (and there will be differences between segments of each of these communities). In addition, communities may differ regarding their preferences for when and/or how a principle may be translated into action, especially where trade-offs must be made.

To provide some flavor of research, education, and library perspectives on these dimensions, this chapter provides a brief overview of some specific principles. The issues raised in this discussion can be addressed by technical, market, and/or policy mechanisms (respectively addressed in Chapters 2, 5, and 6).

Although the discussion below highlights research, education, and library perspectives, resolving key issues will require finding common ground between those perspectives and others relating to industry and/or the general public. The discussions launched by the administration's Information Industry Task Force (IITF); by various interest, professional, and trade organizations; through various electronic discussion groups; and by congressional hearings are all part of the process of weighing and balancing different perspectives.[2] Achieving consensus on fundamental values is an essential element of the process of developing the policy framework for an NII.

EQUITABLE ACCESS

Access to networks and services should be equitable, affordable, and ubiquitous. This principle, expressed in various ways, is perhaps the most universally expressed but also the most elusive in implementation. Success may be difficult to gauge, given the lack of precision or consensus on what constitutes equity, affordability, or ubiquity.[3]

The equitable access principle, its components, and its variants refer to an electronic parallel to telephony's universal service expectation.[4] Some people, for example, are located in areas that are more expensive to serve with any kind of wireline communications. This includes sparsely settled or remote areas, mountainous regions, and other hard-to-reach places. Basic telephone service costs more to provide in these locations than in the cities and suburbs. As a matter of policy, basic telephone service is made available at about the same price regardless of the convenience of location, and the committee believes that the same should be true for the NII.

The experiences in research, education, and library uses of the Internet demonstrate that each community brings new resources and will influence the future direction of the network. *Consequently, efforts and strategies that seek to separate out the communities they serve, especially those that are perceived as most profitable, will serve the public interest less than more inclusive strategies for involvement in information infrastructure.* Moreover, serving the public interest through broad access should eventually serve private interests, by expanding markets and values associated with the infrastructure.[5] This is a key reason for avoiding a situation in which the Internet continues to exist but diverges from the rest of the information infrastructure.

At the most basic level, access to information implies access to infrastructure—access to a network (including associated software) and interconnection of that network to other networks and to resources that can deliver desired information. The nature and degree of access will depend on technical factors (e.g., standards), economic factors (e.g., charges, subsidies, incentives, and so on), facilities, human resource requirements (e.g., training and expertise), motivation, choice of applications and system platforms, and policies (which will influence all of the above). Cost, always a pacing factor in broadening access to new technology, is considered the key barrier to broader access today.

At the next level, access to information raises questions about the mix and range of information services available over the network. Today's research and education communities rely on networks primarily for messaging (electronic mail), accessing information resources and services (e.g., databases), and file-transfer services. In this environment, the users of the infrastructure are the primary producers of the information transferred over it. Tomorrow it is clear that the mix of services will change in two ways. First, the format will change; there will be more video, audio, multimedia, and real-time services, in contrast to the flat text and binary data files that dominate Internet traffic today. Second, there will be more third-party providers of information, which will be offered on a free or at-cost basis by government and on a revenue-generating/profit-making basis by commercial providers.[6] Third-party sources of information (databases, numerical, full-text, bibliographic, weather, news, and other kinds of information) have been in use for many years in libraries, originally on a dial-up basis but most recently via the Internet, and demand has been growing. Of course, they have also been used by various businesses, often on a dial-up basis.

For the research and education communities in particular, an increasing emphasis on network-based access to information raises questions about whether there is a need for a minimum level of access—and what that level is, how basic it is in terms of the quality of service (speed, support for video or multimedia, and so on), the locus of access (on each desk or in each classroom or in such communal and institutional facilities as libraries), and the breadth of reach (within a state or region, national and/or international, connections to which information provider gateways, and so on).

Because some types of research and education are inherently more demanding than others in terms of their bandwidth requirements, equal access to relatively low bandwidth service (e.g., for textual electronic mail) alone would not meet the needs of many users. Although it may be possible, drawing on grant-award and other records, to measure the number of researchers currently funded to undertake high-performance computation and communication, it is much harder to assess how many would do such work were the infrastructure more affordable to them. Similarly, the low-bandwidth/low-budget applications typical of the K-12 experience suggest that the primary and secondary schools have needs that are typically quite modest. However, presented with better infrastructure options, the K-12 demand might well look different—and if given adequate local access capabilities, it could grow dramatically.

In the 1988 report *Toward a National Research Network*, the Computer Science and Telecommunications Board (CSTB) made it clear that there is a basic need for communication that is shared by all and should be available at a price that all can afford:[7]

> To encourage the development and use of a widely accessible network service, the committee suggests that the NRN [National Research Network] might provide (1) a universal basic service at some low to moderate speed for electronic mail at a low cost to users and (2) higher levels of service with an appropriate charging structure. A low charge for basic service would lessen the initial resistance that might otherwise be felt by first-time users and yet discourage the overloading that occurs with free networks. (p. 2)

Although the above statement was formulated with an emphasis on the research community, the essential principles hold more broadly. The present committee endorses the concept of a minimum set of services, as discussed in Chapter 2, that is available to all.

Differences in outlook on equity and ubiquity derive from differences in financial status and ability to pay—hence, the many concerns voiced in the research, education, and library communities about affordability. How network and information access are priced will affect the ability of new users and communities to engage with the information infrastructure.

Linking access and service to ability to pay would be inherently exclusionary. This problem lies behind consumer group and state regulatory efforts to moderate the pace of infrastructure advances and tariff changes. On the other hand, industry efforts to contribute to educational access, in particular, show both awareness in industry of the political appeal of enhancing education and the fact that there are unpredictable alternatives to public financing alone. Although the popular debate over the NII has acknowledged a risk of polarization into information "haves" and "have nots," a more varied spectrum involving "have mores" and "have lesses" may be more realistic.

How ubiquitous is the access will determine, in part, the nature of the benefits: people with desktop or other immediate forms of access will be able to engage in more spontaneous, intermittent, and possibly more interactive uses of the infrastructure than people who must go to a communal point of access and wait their turn. For example, one access point per school represents a far different kind of "ubiquity" than one per teacher or student. Access points for students, in turn, beg the issue of access for parents or guardians, which can increase the impetus for home-based access.[8]

As the issue of home-based access underscores, ubiquity assumes that there is a way of scaling services. To date, there has been a trade-off between effective speed of service and extent of service reach—slower-speed links on the Internet already have become associated with service to less affluent users. While performance on the Internet could be enhanced if access were limited to rates of 56 kbps or above, this might disenfranchise groups (such as schools) that can barely afford to be connected.

From a practical perspective, the key variable is timing. Current administration and congressional activities aim at stimulating or accelerating the achievement of universal access to likely NII services. At the present time there is no compelling argument for a government program to accomplish universal network access to all individuals immediately. Costs would be prohibitive, and the absence of a universal culture of use means that benefits would be swamped by costs. Rather, what is required is assurance that those individuals who desire access can obtain it in a cost-effective and straightforward manner.[9]

The 1988 CSTB report pointed out that the equity issue could be divisive:[10]

> If access is granted on the basis of ability to pay, the network may foster a "rich-get-richer" outcome. . . . [A] situation where large or "rich" users provide their own individual solutions to networking would prevent an NR[E]N from achieving the benefits of scale or access promised by the concept of a truly national network. (p. 24)

This committee concurs with that assessment.

Today, with the move to commercial service, there is the prospect of even greater differentiation than there was in 1988. The committee is pleased to see that the equity principle is captured in the structure and activities of the IITF; the articulation of administration goals for connecting schools, hospitals, libraries, and clinics; and related programs at the National Science Foundation (NSF) and the National Telecommunications and Information Administration (NTIA), including the NTIA Telecommunications and Information Infrastructure Assistance Program that was launched in 1994.

FLOW OF INFORMATION

Access to information through the infrastructure should be governed by rights and responsibilities relating to freedom of expression, intellectual property rights, individual privacy, and data and system security.

Commercial providers of networks and information services and the research, education, and library communities differ in their outlook on the flow of information. While commercial providers focus on the business potential of information—including associated revenue streams—the research, education, and library communities emphasize free sharing of information in the course of their activities, and there is a concomitant emphasis on broad sharing of information.[11] Creation, discovery, and analysis of information are fundamental aspects of research and learning; organization, storage, location, and provision of information are fundamental purposes of libraries; and discovery, communication, analysis, and creation of information are fundamental aspects of education. In the research community in particular, much more information is available and used than is formally published. Rapid communication of data and results over a network are becoming the norm, confounding conventional publishing cycles.[12] The research, education, and library communities have pioneered in the development of electronic library card catalogs, digital library collections, and electronic publications aimed at these communities.

For decades, publishers and commercial information services providers also have been exploring ways to create, produce, and deliver information over networks; their activities have included the development of abstracting services, provision of rapid access to news relevant to businesses, and value-added delivery of legal and financial information aimed at paying clienteles from industry as well as academic markets.[13] More recently, they have perceived a widening popular market.[14] Additional commercial markets relating to network-based delivery of information resources to the educational community are also being explored.[15]

Because of the need to recoup their investments and generate additional capital for service improvements and innovations, controls on access and use are very important to these businesses, as is the ability to charge for access and use.

Funding for the digitization and electronic supply of noncommercial materials is an ongoing concern facing both publishers and consumers of those materials in the research and education communities.[16]

Government Information

The treatment of government-generated information offered through commercial services illustrates various tensions—there has been an ongoing debate in the library, publishing, and consumer advocacy communities about how and by whom information collected or produced by the government should be made available. Some of this debate occurs in conjunction with a broader movement to realize electronic libraries.[17] The research and education communities may have special claims to easy, affordable access to government information, a large amount of which consists of data of interest to—and/or generated by—various kinds of researchers and some of which, in turn, has proven, through activities associated with the National Research and Education Network (NREN) and High Performance Computing and Communications (HPCC) programs, to have a variety of educational applications (Box 4.1).

Such companies as Mead Data Central and WestLaw have made a profitable business of packaging and enhancing legal and other information originating in the government for resale. They have maintained that adding value to information should be done in the private sector, rather than by government, because of the opportunities for promoting competition, the concern that government should not be in the value-adding business (e.g., data interpretation, analysis, and other enhancements), and the concern that government control of the information it generates could result in a monopoly over information made available to the public.[18] Note that government supply does not mean free information; some states, for example, are considering charging for data they supply electronically, in the process assessing the balance between straight cost-recovery and generation of additional revenues and the balance between serving the general public and serving commercial customers.[19]

Public libraries and public interest groups, with the agreement of the information industry, have argued that basic government-generated information should be made available directly, without adding value, to the user, either for free or at the cost of documentation.[20] Given that the government has already incurred the costs of creating and processing the information for governmental purposes, the economic benefit to society

Box 4.1 Government Information

Government programs can stimulate the development of electronic databases of government information that support research and education, and they can support the development of sophisticated data management and searching software that will make the software accessible and usable by the research and education communities.*

Much of the research data available is federally owned and controlled, making the data ideal material for experimental, Internet projects. Furthermore, since the federal data are largely in the public domain, intellectual property problems will not interfere with or confound the research results. Finally, with the increasing use of and reliance on information technologies by federal agencies, it will be important to provide an additional dissemination channel for effective, equitable, and timely access to federal information resources.

Such channels are contemplated by:

- The Clinton-Gore technology initiative, "Technology for America's Economic Growth, a New Direction to Build Economic Strength," which states, "Government information is a public asset. The government will promote the timely and equitable access to government information via a diverse array of sources, both public and private, including state and local government and libraries."
- Recent proposed legislation seeking to stimulate and spur network-based applications of libraries, education, health care, and government information.
- Revised OMB Circular A-130, "The Management of Federal Information Resources," providing guidance to agencies in their collection, maintenance, and dissemination activities and articulating a clear and strong statement in support of public access to government information.
- The Government Printing Office Electronic Information Access Enhancement Act, S. 564 (P.L. 103-40), which moves the Government Printing Office forward into the delivery of on-line federal data resources, including the *Federal Register* and the *Congressional Record.* It calls for the development of an electronic directory of federal public information. These and other on-line data files will be free to depository libraries and available at the incremental cost of dissemination to other users.

---

*This observation is derived in part from discussions among task force members of a joint initiative in the area of intellectual property rights by the Association of American Universities and the Association of Research Libraries.

is maximized when government information is publicly disseminated at the cost of dissemination to both end users and value-added redistributors.[21]

Under the administration's National Performance Review, the NII initiative, and legislative proposals, a number of steps to enhance the flow of government information over the Internet are already being explored. These efforts should be monitored and evaluated in terms of their costs and impact.[22] Internet delivery of government information, like other channels, reaches only a portion of the desired clientele due to various accessibility and affordability barriers, but it provides a laboratory for beginning to assess what works and what does not. In particular, Internet dissemination may help to explore opportunities for reconceptualizing the kind as well as the form and delivery of information of broad public interest, allowing the government, commercial providers, and the public to move beyond the form and content of "government information" based on traditional printed packages.[23]

PRIVACY

Contrasts in outlook between research, education, and libraries and industry have a different flavor in the area of privacy (Box 4.2). On the one hand, there is broad agreement about the need for confidentiality for sensitive information (independent of infrastructure use, per se) about individuals (e.g., information about personnel or personal matters) or organizations (e.g., trade secrets or other proprietary information).[24] Storing and communicating sensitive personal information lead to the possibility that information will be divulged, or that information might be modified, sometimes with life- or welfare-threatening consequences.

On the other hand, information about what consumers do is routinely of interest to businesses, as reflected in the proliferation of frequent-buyer programs, point-of-sale information gathering among retailers, and so on. The collection, analysis, and redistribution of information about individual behavior all have commercial value and have become the basis for a number of commercial database services. For example, market research and demographic data can be used to better target products and produce better or customized products; such data are becoming increasingly valuable. Moreover, individuals will sometimes give up privacy as a consideration for obtaining a service (e.g., medical insurance, credit), although they may not understand the ramifications of that decision.[25]

A principal point of contention appears to be information about patterns of network and information service use, information that is likely to be collected routinely through accounting mechanisms associated with use (and especially purchase) of network and information services. Such

Box 4.2 Confidentiality in Context: Access Control Needs

Community:	Key Rationale:
Research	• Avoid being scooped • Avoid mischief and preserve integrity of databases and research writeups • Protect individual researcher privacy • Avoid giving away market research • Prevent competitive advantage in grant-seeking derived from knowledge of researcher's patterns of information use
Education	• Protect privacy of student records • Protect individual educator privacy, privacy of educator interactions • Prevent unauthorized disclosure of test and homework answers • Avoid mischief and preserve integrity of educational materials • Prevent unauthorized access to sensitive material (e.g., pornography) while preserving First Amendment rights • Ensure that intellectual property rights requirements are met
Libraries	• Protect privacy of library user records (e.g., who borrowed or accessed what)
Hospitals/health	• Protect privacy of patient records • Protect privacy of physician consultations, records • Avoid mischief and preserve integrity of medical records, test results, and so on • Protect eventual flows of funds

information collection has been an issue in the deregulation of telephony, where the debates have revolved around "customer proprietary network information" and "automatic number identification" information; these debates have begun to be carried forward into considerations of the larger NII policy.[26] Electronic networking makes possible more frequent and more extensive collection of user behavior data as an automatic by-product of use, and it facilitates the amassing of data from multiple sources accessible over networks. As a result, privacy implications of such data gathering and analysis are attracting broader attention.[27] *The treatment of information about infrastructure use—in particular, who owns and accesses what information—is an issue that needs further exploration and resolution,* as has been recognized through the organization of the administration's IITF.[28]

A central issue with regard to protecting privacy is the balancing of legitimate business opportunities against individual rights, which need to be more clearly defined in this context. As Roger Noll explains,

> Policy regarding privacy faces the classic trade-off in values that is common to all forms of information. A right to privacy enhances the value of personal information to individuals, but dissemination of the information without compensation provides value to others. . . . The policy dilemma, then, is a fundamental conflict over which values count the most. . . .[29]

The research community is not universally aware of or concerned about privacy, but it has included individuals who have been particularly vocal in raising such concerns, in part because their experience with networking has underscored the risks.[30] The library community has a tradition of protecting the confidentiality of borrowing records and has worked successfully to provide legal protection for the borrowing community. For example, many states have laws prohibiting the release of information about what an individual is reading, watching, or listening to (this restriction could apply to video rentals and other information access activities). (Within the education community, the greatest concern about privacy may relate to the confidentiality of student records, which generally is protected under law.) Within the medical community, the need to protect the privacy of patient records is attracting growing recognition.[31] Given this experience base, efforts to assess and decide on privacy policy (such as those undertaken under the IITF aegis) should explicitly involve participation from the research, education, and library communities.

FIRST AMENDMENT

There is general agreement that the First Amendment, which fosters the broad dissemination of divergent viewpoints, applies as much to the electronic, multimedia environment as it does to the print environment. *Translated into a principle for information infrastructure, the First Amendment suggests that government should permit no one to exercise monopoly control over the content carried over the network; content determination and editorial control issues should be the province of competing information providers.*[32] Monopolization of content is a fear that arises in the context of vertical integration (combined provision of content and conduit, as in the case of cable and broadcast television, which are regulated in different ways to protect public interests in content). A related issue is placement on the network and in address directories, which ideally should be provided without discrimination or favoritism. It is in the public interest to have a variety of information sources.

An important lesson from the Internet experience is that users should have the ability to place information on the network—a capability that has differentiated the Internet, whose users have emerged as both authors and publishers. A corollary is a right of users to offer services over the network.[33] How to extend this feature into a larger, commercial environment raises many questions, since the First Amendment provides for freedom to publish by media owners, but it does not necessarily provide a right of access to media. What the Internet experience shows, however, is that the convergence of digital media enables an effective separation of content from the carrier. People are increasingly able to create, deliver, and express in text, images, and sound regardless of the medium—one can define an information product by what it does rather than by the medium to which it is bound; there may not be a physical medium.[34]

The Internet experience also illustrates the value of broad-based, many-to-many expression, in contrast to telephony, which is generally point-to-point communication, which inherently limits the scope of one's audience. The Internet provides individuals with access to mass communication channels, a feature particularly evident in its use during popular uprisings in the People's Republic of China and the former Soviet Union.

However, open interfaces and nondiscriminatory access to media can produce outcomes that trouble some for various reasons (such as 900-number pornography services in the case of telephony); some of these are subject to legal protections, and others are expected to be addressed by current information policy efforts. In K-12 education, content restrictions are driven by the presence of minors and varying sensitivities about how much access children should have to "adult" material of various kinds. To date, school networking projects have acted to restrict access to pornographic and other controversial information available via the Internet (which is not necessarily possible) and also to educate children, teachers, and parents about appropriate materials.[35] Exploration continues. It is worth noting that libraries have a large base of experience in this domain, having fought—usually successfully—censorship battles in every information format introduced to date. Librarians have developed both policies and tools and put them into regular practice.

Within the NREN environment, the scope of expression has been explicitly constrained. The most notable attempt to do so is through acceptable use policies that have arisen in part to control how government-funded networks are used. For example, NSF has imposed an honor-code acceptable use policy (AUP) that has restricted the content of communication over the NSFNET and militated against the conduct of commercial transactions over the Internet; similarly, research networks established by the Department of Energy and the National Aeronautics and Space Administration (NASA) limit traffic to mission-related transactions. Although AUPs have caused controversy (from allegations of censorship to observations that they are impossible to police), it is presumed that they will diminish with the decline in federal support for network infrastructure.[36] That is, the federally supported networks with AUPs, such as the anticipated NSF very high speed backbone network service (vBNS), will constitute a shrinking portion of the Internet (and NII). The current NSF cooperative agreement establishing network access points will allow the AUP-constrained vBNS to be connected to networks the rest of whose traffic is free of the NSF AUP.

During times of technological change, the interpretation of constitutional protections, such as the First Amendment, is normally developed by a long litigation process with case law being developed and further interpreted. Over the past 200 years this model has served us well primarily due to the relative slowness of technological change. The environment to be created by the NII suggests that the models implicit in case law may need to be revisited. A number of questions need to be addressed, for example, What is a bulletin board operator—a publisher, a book store owner, or a new type of enterprise? and, What are its liabilities and responsibilities with respect to the information it is custodian of? What are the responsibilities of the new unregulated data carriers, including e-mail service providers, with respect to the material they carry? The speed with which these issues are faced will affect the growth of the NII culture. Decisions that are excessively restrictive may slow it down or change its nature. A balance must be found between the rights of the operators and those of the users.[37]

INTELLECTUAL PROPERTY PROTECTION

There is a broad public consensus that there should continue to be protection of copyrighted materials, and that the protection of copyrights of materials distributed in the networked environment should reflect the rights of both creators and users.[38] There is also broad appreciation that a robust market for networked information and resources is fundamental to the success of the evolving NII. Much less certain is how intellectual property protection can or should evolve to fit the networked environment. Observed legal scholar Thomas Dreier,[39]

> How do digital technology and networks affect [the] background against which copyright was previously applied? Obviously, digital technology alone doesn't, since a work stored on a material carrier in digital form

> also travels the route from private via public to private. Rather, the problem caused by digital technology as such is one concerning the ease of reproduction. However, it is the network which brings about the substantive change. It links the private sphere of the author—or of the person or entity offering the work in its marketable form—directly to the private sphere of the person who enjoys or re-uses the work. Thus, the public sphere on which copyright relies to such a great extent is eliminated, and little more is left than the umbilical cord of the connecting net-line which runs through what used to be the now-eliminated former public sphere. It also follows from the disappearance of the public sphere that any person enjoying or re-using a protected work via a network, reaches from his or her own private sphere directly into the private sphere of the author who makes the protected work available. (p. 193)

Intellectual property in the networked environment is bound up with how intellectual property protection should evolve to cover digital media generally. For example, copyright law applies to the multimedia environment as it applies to the print world, but some have raised questions about the application of intellectual property protections to the hybrid products made possible by digital convergence. How these questions are resolved will affect business development and process planning.

The area of intellectual property rights represents one of the greatest areas of difference among the research and education and commercial communities. To begin with, it is worth noting that there is broad agreement on one aspect of intellectual property protection: the importance of protecting data integrity. It is critical that creators know that what they produce is what network users get, and for users to be assured that what they are getting is what they think it is. The research, education, and library perspective was articulated by the University of California's Clifford Lynch as follows:[40]

> In discussing issues of access and integrity in networked information today, there is a very strong bias towards issues specific to scholarly information; this is to be expected, given that the academic and research communities have up until now been the primary constituencies on the Internet. These are relatively sophisticated communities, and communities with an ethos that is strongly oriented towards citation, attribution, and preservation of the scholarly record. Indeed, . . . this ethos ties these communities to the system of print publication, and emphasizes that networked information must offer the same integrity and access if it is to become an acceptable replacement for print. . . . (p. 16)

Concerns relating to integrity suggest a need to protect against the dissemination of an altered original work without authorization, or if this is not possible, a means for "certifying" the authenticity of the data.[41] (Mechanisms to do this may be developed under the security umbrella; see Chapter 2.)

Where there is fundamental disagreement about intellectual property protection is in the area of charging for both acquisition of the item itself and then for each use, since property rights—a form of ownership—carry with them an inherent ability to reap value through sales. Charging for copyrighted materials delivered over the Internet was contemplated in the High-Performance Computing Act of 1991.[42] The opportunity for financial compensation is the reason that intellectual property protection, especially copyright, is broadly recognized as nurturing the quantity and diversity of information creation generally in the commercial marketplace.

In an electronic environment, the potential for violation of the law (and for related theories in the law such as misappropriation) is greatly exacerbated. It is increasingly possible for virtually anyone to make a perfect copy of an information product. Indeed, the ability to make multiple perfect digital copies, accessible instantly to millions of users worldwide, is an exciting prospect, but it can also be a license to steal intellectual property cheaply, easily, and in a way that destroys incentives for future creativity. This battle has already been fought by the software industry, which has persuaded at least large organizations to buy or license appropriate numbers of copies—their educational and legal efforts attest to the possibilities for (re)educating the public.

Recognizing that the Internet has grown up within a culture that thrives on sharing information and research, commercial information providers have approached the network with great caution. Commercial authors and publishers are concerned that their works will be accessed, easily downloaded, and redistributed without appropriate authorization and remuneration.[43] The scale of a network such as the Internet makes this prospect a substantial deterrent to making information available. Publishers and commercial information service providers believe that if this issue is not adequately addressed, not only will many information providers be unwilling to trust their assets to these networks, but they will also, in fact, have a strong incentive to keep their property from ever being included. This obviously runs counter to the most central goals of the NREN and NII efforts.

In the research community, the relatively low emphasis on remuneration is one factor behind a greater willingness to publish electronically. Others include a need for rapid dissemination of information and broad circulation of new ideas, "food for thought," and research results.[44] These factors have given rise to a substantial amount of experimentation with and demand for electronic publishing capabilities, services, and support.[45]

Universities and their faculties, students, and libraries play multiple roles in relation to intellectual property: as creators, users, disseminators, and archivers. U.S. research and education have a special place in copyright law through provisions for fair use, libraries, archives, classroom use, and some performances and displays. While universities eagerly lay claim to patents (revenue-producing), they often forego any claim on copyrights (cost-saving), and faculty often readily transfer copyrights to publishers in exchange for achieving publication and the chance for published knowledge to be attributed to them.

The end result is that universities sometimes find themselves in a situation where they cannot afford to buy back—through serials and other acquisitions—the research information their faculty have generated in their service. In addition, the campus environment is one of a general lack of knowledge about and understanding of the current copyright law, its fair use provisions, and the ramifications of electronic publishing. Where these issues are being addressed, several options are being considered (Box 4.3).

In contrast to the academic situation, there are more formal transactions involving rights and remuneration in the commercial publishing arena. However, electronic publication has raised new questions in that arena that may affect all kinds of authors. Specifically, authors may be less knowledgeable than publishers about options for republication and associated revenues through electronic databases and other electronic vehicles.[46] Thus, the author may focus on the original publication, and not a far broader publishing process through electronic databases.

Overall, however, authors as well as publishers are experimenting in the commercialization of these paths and in the charging for and control of the intellectual material flowing and stored in electronic form. Publishers are experimenting with new ways to distribute information and manage copyright over electronic networks; they are experimenting with developing copyright management mechanisms using digital technology and aimed at detecting, monitoring, and inhibiting improper use.[47] Some of those experiments revolve around the academic community. For example, the Elsevier TULIP project, involving several U.S. universities, is exploring technical and economic issues relating to the use of campus networks and the Internet to distribute journals. Springer-Verlag has an experimental project (Red Sage) based on a Bell Laboratories system for delivery of content from 24 journals over the Internet. It can manage journal page image content on a network and alert users by electronic mail when subjects that interest them are covered in journals they do not ordinarily read.[48]

The possibilities for republication raise questions about shifts in the costs and benefits of publishing.[49] Efficient mechanisms for negotiating and paying for partial or complete copying, for "reproduction" rights, and so on are still in development (they exist in certain closed electronic

Box 4.3 Copyright Management Options Under University Policy and Support

Faculty Ownership. The faculty member, as the author under the copyright law, retains copyright. The faculty member is then free to license uses within the university (and higher education) or assigns or licenses to the publisher the right to reproduce and distribute the work as necessary.

Joint Faculty/University Ownership. The faculty member and the university share jointly rights in copyrighted works. Such joint ownership would be limited to that work generated within the scope of the faculty member's employment.

University Ownership. The university, as copyright holder, makes all decisions relative to publication, licensing, royalty agreements, and so on. Rights are transferred to the university for works generated within the scope of the faculty member's employment.

Ownership by a Consortial Body. The faculty member assigns rights to a consortial body, e.g., the National Association of State Universities and Land Grant Colleges or the American Association of Universities, which then acts as a collective rights society to manage copyrighted works in the best interests of universities and their faculties.

delivery systems, such as for the NEXIS and Dialog services), even though technology provides ultralow cost mechanisms for reproduction and copying. Early in 1994, for example, Dialog Information Systems announced its intention to adopt a charging system based on the number of authorized users, effectively selling the right to share electronically the documents obtained through its system. Although such an approach would, if enforceable, reduce loss of revenue from unauthorized copying, it adds to the growing body of questions about redistribution and reprint rights.[50]

Experimentation within the research and education communities suggests that in the future, the modes of information consumption may change or there may be greater variety in modes. For example, there is a tendency toward more casual publishing growing out of liberal electronic mail, public file repository, Gopher server, and Mosaic/World-Wide Web experiences. (See Box 2.2 in Chapter 2.)

There also appear to be more "synthetic" applications that involve using small amounts of material from multiple sources, and more short-lived uses of information (in some cases because the information is frequently updated, in other cases as interactive databases or other information resources proliferate). So-called "hot links" allow users to reach out from one kind of document into other, related material through the network; the Mosaic interface to the World-Wide Web illustrates this phenomenon. In addition, markets are demanding products by the piece (e.g., single articles) rather than the conventional publishing package, and pieces are being integrated by users.

The combination of new applications and a broadening user base challenge conventional approaches to pricing for information services. For example, current expectations about amount of royalty per sale or use may be inappropriate in an environment where a far larger number of people may seek to use the information and/or where the uses may be more short-lived than has been customary with paper or other fixed-medium products. Overall, all players in the NII publishing area must reevaluate their economic expectations and adapt them to the realities of the electronic world.

BROADER CONSIDERATION OF ETHICS

The committee anticipates that the emerging NII will dramatically increase confrontations among conflicting rights and needs of society. These conflicts will not necessarily be about new issues, but rather about a change in the magnitude and urgency of those that already exist. Core issues will revolve around ability to access information, the cost of that access, and respect for intellectual property, privacy, security, and information integrity. Triggering conditions may include the increased ability to collect and collate data on all aspects of our lives and the increased ability to cause harm by altering public and private records. Technological fixes to these problems are not in sight. The ethical, legal, and social implications of this technology are paralleled by those arising from discoveries in molecular biology. Some of these issues will probably be worked out in the marketplace, some in the courts. As in other domains, government monitoring and possibly more active involvement may become necessary. More important, the courts unquestionably will need guidance in issues that demand a high level of technological knowledge, but are not simply technological problems with technological fixes. And there will certainly be a need to increase public awareness of many complex issues.

Building from the IITF information policy efforts and depending on their results, consideration should be given to a program to fund research and public education in this area to provide the guidance that will be needed. A reasonable model would be the Ethical, Legal, and Social Implications (ELSI) program of the Human Genome Project.[51] The goals of an ELSI component to the NII would be as follows:[52]

1. Address and anticipate the implications for individuals and society of the NII;

2. Develop policy options to assure that the information is used for the benefit of the individuals and society.

3. Enhance and expand public and professional education that is sensitive to individual rights and responsibilities associated with the NII.

Examples of such efforts would be workshops, panels to frame policy options, training grants for those with a technological background to acquire training in ethics and law (and vice versa), studies about how current opportunities for abuse are being exploited, public education programs, and pilot programs for introducing the study of such issues in the schools. Within NSF's Computer and Information Science and Engineering Directorate, there are programs (e.g., within the networking and cross-disciplinary activities divisions) that could house such a research program, although the scope of the problem may dictate a more broadly based institutional home, such as a national commission.

NOTES

1. This development has been anticipated by the incorporation of activities relating to information policy under the administration's Information Infrastructure Task Force as well as by the corresponding broadening of technical activities under the High-Performance Computing and Communications umbrella.

2. See, for example, U.S. Department of Commerce. 1994. *20/20 Vision: The Development of a National Information Infrastructure*. Government Printing Office, Washington, D.C., March. This volume contains papers presented at a special conference that was requested by the administration to stimulate debate over the broad principles and goals advanced in the fall 1993 policy statement on the NII (Information Infrastructure Task Force. 1993. *The National Information Infrastructure: Agenda for Action*. Washington, D.C., September 15).

3. The issue of metrics and evaluation is recognized by the administration, which posed the problem to speakers at its March 1994 20/20 Vision conference.

4. Although the analogy is frequently made today to universal telephone service, it should also be recognized that that entitlement, to the extent it is one, does not extend to universal access to 900-number or other special services, nor are newspaper publishers obligated to distribute free or discounted copies to the indigent—who can and do read the paper in public libraries. Also, recognizing that today's universal access for voice telephony took time to be implemented, it is likely that in some transition period there will be inequalities.

5. For example, this tack has been taken by the electronic text archive developers at the University of Virginia, who hope that their apparent success will stimulate commercial publishers. See Seaman, David. 1993. "Gate-keeping a Garden of Etext Delights: Electronic Texts and the Humanities at the University of Virginia Library." *Gateways, Gatekeepers, and Roles in the Information Omniverse: Proceedings from the Third Symposium*, November 13-15, Washington, D.C., pp. 63-67.

6. Of course, continued "publishing" by information-producing individuals is still expected and desirable.

7. Computer Science and Technology Board (CSTB), National Research Council. 1988.

Toward a National Research Network. National Academy Press, Washington, D.C. (The Computer Science and Technology Board became the Computer Science and Telecommunications Board in 1990.)

8. A significant percentage of the population is now network literate and has a computer at home. According to the Department of Commerce, over 31 million U.S. households owned personal computers in 1993; two-thirds of home PCs are believed to support people who work from their homes. See U.S. Department of Commerce. 1994. *U.S. Industrial Outlook 1994.* Government Printing Office, Washington, D.C.

While the home-working, network-literate sector of the population will probably never be sufficient to motivate the universal, high-bandwidth networking of the country (which might, however, be motivated by entertainment and consumer activities), it is a sector that must be assured reasonable access to the Open Data Network, for reasons of public policy such as balanced access to network information. Today, it is very difficult for individuals to procure this service, to some extent for reasons of cost but more basically because there is no standard way of purchasing the service.

General consumer access does not appear to justify any special subsidy. The consumer should be asked to pay for attachment, assuming that the cost will be reasonable. What is required is a program to develop a well-defined means to provide this service. For example, if a specific initiative were undertaken by the government to drive the deployment of low-cost networking into schools and public libraries, it could promote the development of technology suited for the network attachment of individuals (assuming the small number of schools per central office did not diminish the impact and appeal).

9. Existing and likely demand levels present a natural pacing factor, but it should be recognized that there may be an exaggerated lag in the development of desire for access. In particular, educators, for example, in schools in unaffluent areas and educators without exposure and training relating to the potential of networks and information resources may not know enough to desire access.

10. CSTB, 1988, *Toward a National Research Network.*

11. Traditionally, some sharing has been limited as a function of competition for tenure and credit, but this resistance has been diminishing in the face of pressures promoting collaboration.

12. Electronic publishing will change many dynamics affecting the scholarly record, the dissemination of research results, and the evaluation of the work of researchers. Some of the confusion now arising is reflected in the following comments (made at a Modern Language Association Convention in Toronto on December 29, 1993) of Bill Readings of the University of Montreal: "If an article is never technically 'in print,' how do you know it was actually 'published'? If revisions and new footnotes and retractions and formal responses become parts of the electronic archive, doesn't 'publication' cease to mean a finished artifact and become instead an unending dialogue of readers and authors? What then becomes of journal editors and their screening panels, the so-called 'gatekeepers' of scholarship, when space and cost considerations vanish and you only need one reader to legitimate publication on the Internet? Many accuse academics of only writing for themselves. . . . The virtual university gives new meaning to the quip." See Trueheart, Charles. 1993. "The Writing on the Cyberspace Chalkboard," *Washington Post,* December 30, pp. C1-2.

13. A fundamental issue in relation to acquiring, storing, and sharing information resources in the electronic environment is ownership—who owns the information has had an enormous impact on the prices charged to the customer. A number of cost-per-unit studies of various kinds of scientific, technical, and medical (STM) journal pricing suggest that the not-for-profit producers sell information significantly less than the for-profit publishers, specifically the large, offshore European STM publishers. See Okerson, Ann. 1993. "Networked Information, Scholarly Publishing, and Electronic Resource Sharing in Academic

Libraries: a Dilemma of Ownership," paper distributed by Association of Research Libraries, Washington, D.C., June.

14. Within one week alone in early 1994, there were several announcements of new services, including a Los Angeles Times-Pacific Telesis electronic shopping service; the launch of a TV evening newscast by the *Philadelphia Inquirer*, and the formation of an alliance to develop an electronic publishing service by the *Washington Post* and Oracle. Glaberson, William. 1994. "Newspapers Race for Outlets in Electronic Marketplace," *New York Times*, n.d., pp. D1 and D6.

15. For example, Encyclopedia Britannica Inc. recently announced plans for electronic distribution to universities and some public libraries over the Internet. Of the several encyclopedias already available on-line, *Encyclopedia Britannica* is the largest. The Britannica Online service being tested at the University of California at San Diego features hypertext links that cannot be matched by traditional text versions; links among the macropedia, micropedia, index, and propedia broad outline of world knowledge; and WAIS and Mosaic access. At the time of the announcement, the pricing was uncertain—options for pricing on a subscription basis and reference by reference (enabled by WAIS and Mosaic) were being considered. See Markoff, John. 1994. "Britannica's 44 Million Words Are Going On Line," *New York Times*, February 8, pp. D1-2.

16. Participants in a 1993 workshop noted that there are many questions about the division of labor between public and private sectors in creating, storing, organizing, marketing, and distributing noncommercial information. See Library of Congress. n.d. "Delivering Electronic Information in a Knowledge-based Democracy," Summary of Conference Proceedings, July 14, 1993.

17. The High Performance Computing Act of 1991 and other pieces of legislation have increased support for and exploration of electronic libraries.

18. Some of these arguments are reflected in the Paperwork Reduction Act (PL 96-511) and Office of Management and Budget (OMB) Circular A-130 (revised in 1993) on federal information resources management. The Information Industries Association has published a pamphlet entitled "Serving Citizens in the Information Age: Access Principles for State and Local Government Information," which provides guidelines in support of "three fundamental tenets": a broad public right of access, a right of nondiscriminatory access, and a prohibition on government control of information access and use.

19. Rohter, Larry. 1994. "Florida Weighs Fees for Its Computer Data: Some See Profits; Others, Too High a Price," *New York Times*, March 31.

20. See, for example, Bass, Brad. 1993. "EDGAR/Internet Effort Leads Public Access to Federal Data," *Federal Computer Week*, November 1, p. 9.

21. One relevant interagency proposal is for a Government Information Locator System (GILS) describing each agency system containing publicly accessible information plus other cataloging information. The proposal contemplated a "wholesale" service, as opposed to direct end-user access. Power, Kevin. 1993. "How Will We Navigate the Federal Data Maze?" *Government Computer News*, November 22, p. 47. The OMB plans to issue a directive establishing the GILS and outlining what agencies have to do to comply with it, including creation of agency inventories of information dissemination products and automated information systems. Sikorovsky, Elizabeth. 1994. "OMB to Issue Guide to Establishing a Governmentwide Locator System," *Federal Computer Week*, April 25, p. 6.

22. For an assessment of options, programs to date, and dimensions on which evaluations might be conducted, see U.S. Congress, Office of Technology Assessment. 1993. *Making Government Work: Electronic Delivery of Federal Services*. OTA-TCT-578. Government Printing Office, Washington, D.C., September.

23. Fisher, Francis Dummer. 1994. "Open Sesame! How to Get to the Treasure of Electronic Information." In *20/20 Vision: The Development of a National Information Infrastructure*.

U.S. Department of Commerce, National Telecommunications and Information Administration.

24. According to a 1993 Louis Harris survey, 53 percent of Americans are very concerned about their privacy. That survey marked the first time such concern was registered by a majority in 23 years. Privacy concerns were greatest for dealings with banks, health insurers, and hospitals among various types of businesses. See "Mind Your Own Business," *Wall Street Journal*, October 5, 1993.

25. A significant feature of the draft "Principles for Providing and Using Personal Information" and their "Commentary," distributed for comments by the Information Infrastructure Task Force Information Policy Committee Working Group on Privacy in the NII on May 4, 1994, is the "protection principle," which states that "Users of personal information must take reasonable steps to prevent the information they have from being disclosed or altered improperly. Such users should: (1) Use appropriate managerial and technical controls to protect the confidentiality and integrity of personal information" (p. 3). Thus, although the IITF also provides an education principle, recognizing that individuals may not understand the impacts of networked information, it argues that individuals should take an active responsibility for information concerning themselves. See Appendix B.

26. See, for example, "Privacy Comments Urge Industry Self-Regulation." 1994. *Telecommunications Reports*, April 4, pp. 32-3, which documents comments submitted in NTIA proceedings relating to privacy and the NII. Telephone companies have argued against additional regulation; the National Cable Television Association and others have called for technology-neutral approaches; distinctions have been posed between actions relating to data misuse versus actions relating to collection, transmission, and analysis of data, per se.

27. Various frequent buyer programs, including the original frequent flyer programs started by airlines and more recently grocery store-based systems, are examples.

28. There is also the prospect of legislation to prohibit the association of the user's name with the specific user data themselves.

29. Noll, Roger G. 1993. "The Economics of Information: A User's Guide." Pp. 25-52 in *The Knowledge Economy: The Nature of Information in the 21st Century*. Institute for Information Studies, Nashville, Tenn.

30. For example, the computing and communications community has organized an electronic forum, RISKS, devoted to sharing information about and discussing a variety of risks relating to privacy, security, safety, and the reliability of computer-based systems. There are other electronic discussion groups specifically focused on privacy.

31. Institute of Medicine. 1991. *The Computer-Based Patient Record*. National Academy Press, Washington, D.C.; and Institute of Medicine. 1994. *Health Data in the Information Age: Use, Disclosure, and Privacy*. National Academy Press, Washington, D.C.

32. The Taxpayer Assets Project, for example, sent a letter to Congressman Markey expressing concern about increasing vertical integration in service provision, inasmuch as telephone companies, cable companies, and wireless carriers can own rights to programming content available through their facilities (November 10, 1993, e-mail).

33. However, most users who want to offer information will not necessarily want to make the commitment involved in offering a service (e.g., running a computer in their homes with the reliability expected from a service provider). New services providing disk storage and reliable access might emerge as information intermediaries.

34. Baron, David. 1993. "Digital Technology and the Implications for Intellectual Property." Pp. 29-36 in *WIPO Worldwide Symposium on the Impact of Digital Technology on Copyright and Neighboring Rights*. WIPO, Geneva.

35. Schwartz, John. 1993. "Caution: Children at Play on Information Highway," *Washington Post*, pp. A1 and A26.

36. This trend is evident in the anticipated proliferation of network access points in the new NSFNET architecture.

37. The Computer Science and Telecommunications Board is addressing some of these issues through its project "Rights and Responsibilities for Participants in Networked Communities," for which a report is anticipated in mid-1994.

38. The Copyright Act strikes such a balance and in particular, section 107 stipulates fair use provisions that should be considered in a networked environment. It is too early to consider or to impose restrictive agreements that limit access, limit creativity, or undermine fair use provisions as these could be detrimental to long-standing policies that promote public access.

39. Dreier, Thomas K. 1993. "Copyright Digitized: Philosophical Impacts and Practical Implications for Information Exchange in Digital Networks." Pp. 187-212 in *WIPO Worldwide Symposium on the Impact of Digital Technology on Copyright and Neighboring Rights*. WIPO, Geneva.

40. Lynch, Clifford A. 1993. *Accessibility and Integrity of Networked Information Collections.* Background Report/Contractor Report prepared for the Office of Technology Assessment, July 5.

41. Imagine the disasters that could occur if, for example, chemistry laboratory manuals were tampered with. Consider the liability issues. Consumers should have reason to believe that the procedures for conducting a potentially explosive experiment, for example, are accurate to the best of the chemist/author's knowledge and have not been altered by someone using a home chemistry set. Users must be able to choose sources of information with a reasonable degree of confidence.

42. In the interest of promoting the development of information services, such as federal scientific data, directories, and commercial information services in support of research and education, "[t]he Network shall provide access, to the extent practicable, to electronic information resources maintained by libraries, research facilities, publishers, and affiliated organizations . . . [and] have accounting mechanisms which allow users or groups of users to be charged for their usage of copyrighted materials available over the Network and, where appropriate and technically feasible, for their usage of the Network" (PL 102-194, section 102).

43. This issue arises with other modes of electronic dissemination. For example, the availability of CD-ROM publications at the Library of Congress was achieved only after two years of negotiations with publishing and library entities; physical security and access restrictions were part of the agreement. See Minahan, Tim. 1992. "With Copyright Protections in Place, Library Receives First CD-ROM Publications," *Government Computer News*, November 22, p. 10.

44. For example, James Noblitt, head of the Institute for Academic Technology at the University of North Carolina, was recently quoted as saying, "These days the world changes every six months and textbooks come out every three or four years. The technology is now in place to keep up with change, but textbooks can't do it." Cox, Meg. 1993. "Technology Threatens to Shatter the World of College Textbooks," *Wall Street Journal*, June 1, pp. A1 and A5.

45. Perritt, Henry H., Jr. 1992. "Market Structures for Electronic Publishing and Electronic Contracting on a National Research and Education Network: Defining Added Value." *Building Information Infrastructure: The NREN and the Market*. Harvard University Press, Cambridge, Mass.

46. Bald, Margaret. 1993. "The Case of the Disappearing Author," *Serials Review* 19 (3, Fall):7-14.

47. There are a variety of joint university-industry projects aimed at developing technological means of providing on-line authorization, tracking, and/or collection of royalties. However, the results of these efforts are treated as proprietary information. In addition, the Corporation for National Research Initiatives has been working with the Library of Congress/Copyright Office and ARPA on network-based copyright management.

See also Oman, Ralph. 1993. "Reflections on Digital Technology: 'The Shape of Things to Come.'" Pp. 21-24 in *WIPO Worldwide Symposium on the Impact of Digital Technology on Copyright and Neighboring Rights*. WIPO, Geneva.

48. "Electronic Publishing Case Studies: Implications for Periodicals Publishing," *PSP Bulletin* 7(3, Winter 1993):3-5.

49. See, for example, "Writer Sues Publishers Over Articles in Databases," *Wall Street Journal*, Friday, December 17, 1993, p. B5(w).

50. Riordan, Teresa. 1994. "Fee Plan to Share On-Line Data," *New York Times*, April 6, pp. D1 and D6.

51. The idea was first suggested to the committee in a June 1993 briefing by Richard Civille, Center for Civic Networking Groups, and later explored and expanded by biologist and committee member Charles Taylor.

52. The Human Genome Project initially budgeted 3 percent of its budget for its ELSI component and has increased this to 5 percent for the next 5-year period. The committee specifically refrains from making recommendations about funding amounts at this time, recognizing that in its inherent focus on the nature of life the Human Genome Project presents more uncomfortable ethical and policy challenges than the NII, but the portion chosen should be a correspondingly significant part of the NII budget.

5

Financial Issues

The evolution of the Internet has several financial dimensions. These include, in the context of this report, the changing role, targeting, stability, and level of federal funding to support the supply and use of information infrastructure, as well as the changing exposure of Internet users in the research and education communities to infrastructure expenses on the demand side. As federal infrastructure activities expand from a National Research and Education Network (NREN) to a diffuse and private-sector-dominated National Information Infrastructure (NII) program, the research and education communities present both specific policy concerns and insights into the larger problem of financing access across the population.

FEDERAL AND OTHER FUNDING FOR NETWORKING TO DATE

Federal financing for information infrastructure fits into a large complex of decisions: choices are made about spending on research and education versus other activities, on different kinds of research, on different kinds of education, and on elements that enter into research and education as inputs, such as networks. Of all these choices, the federal role in supporting research in general is perhaps the most established. Decision making associated with the federal research and development budget has placed a high value on both the development and use of networks to support the research community, in particular. Federal support has yielded enormous benefits: increased understanding of how networks and network-based resources can be used, enhanced efficiency and effectiveness of research through easier sharing and communication, and qualitative change in the nature of research undertaken. These are hard-to-quantify benefits that must be compared to the more obvious costs. However, the difficulty of making that comparison in practice inhibits the adoption of improved information technologies.[1]

In the coming months direct financial support by the federal government for the Internet will begin to diminish.[2] Key developments are the shift to commercial service anticipated in 1994 for what has been the NSFNET backbone and, over the next four years, the National Science Foundation (NSF)-supported regional networks (see Box 1.3 in Chapter 1). Thus we find the uncomfortable situation that those parts of the research and education community that have benefited most from NSFNET face the prospect of changing to a new mode of payment for network-based services while presumably maintaining current commitments and continuing to contribute to the expanding information infrastructure. Although the actual amount of federal support is smaller than most in the research community (in particular) believe, the NSFNET transition has prompted considerable uncertainty and anxiety. These concerns are a legacy of federal Internet support, which has created a dependence on the Internet.

Meanwhile, new federal programs (such as the Telecommunications and Information Infrastructure Assistance Program of the National Telecommunications and Information Administration (NTIA)) and new industry efforts (such as the Pacific Bell plan to wire up schools in its service areas) will result in further redistribution of the resources supporting the use of information infrastructure. Pacific Bell announced in February 1994 a $100 million program for "linkage for computer communications and videoconferencing" to nearly 7,400 public K-12 schools, public libraries, and community colleges in its service area by the end of 1996.[3] As a result of these developments, some communities that have found network access difficult, such as the K-12 schools and hospitals, should gain greater access to network-based information resources and services generally and advance their use of the Internet in particular. However, actual amounts of new or redirected federal funds may be relatively limited. For example, in a February 1994 speech, President Clinton affirmed his support for extending connectivity to libraries, but added that he was "afraid that the lion's share of that work will have to be done at the state and local level."[4]

To the extent that redistribution of federal funds helps others to join the network without overconstraining the participation of incumbent users, benefits will be experienced more broadly. (The alternative, of course, is a zero-sum situation in which early beneficiaries, such as certain segments of the research community, suffer uncompensated reductions in support so that new or inadequately supported beneficiaries can gain support.) The broader distribution of benefits is a stated goal of current NII-related initiatives. Realizing that goal requires the integration of the NREN objectives, which were built on government funding, with the mission of commercial enterprises, which presumes private funding.

Funds associated with the NREN program have been allocated for three purposes: to construct and grow the infrastructure, to pay the ongoing costs of network operation, and to underwrite fundamental research and development.[5] Box 5.1 presents relevant federal funding figures. The federal government has played a leadership role in the Internet environment by contributing funds for development of technologies from protocols to routers, and for construction of such components as the NSFNET backbone, mission agency networks (e.g., ESnet, NASA Science Internet), and the Federal Internet Exchanges. Network construction investments, in particular, have been small from a national perspective (tens of millions of dollars), in comparison to investments for other infrastructure elements, including telephone and cable networks, that have been privately financed (tens to hundreds of billions of dollars). The transition to commercial Internet service shifts attention from federal investment in network construction to support for network operation, maintenance, and use. See "Influence on Network Deployment" in Chapter 6.

The establishment of regional networks in cooperation with state governments and industry illustrates how the federal government leveraged a relatively small investment for the benefit of a large and dispersed community. The NSF could not afford to finance broad, direct institutional connectivity to supercomputing centers, its initial research networking concern, and so it catalyzed a set of midlevel (i.e., between the intra-institutional and national levels) networks that was charged with developing other sources of support over time (see Appendix A).[6] Today there are several regional networks that individually receive the equivalent of $3,000 to $6,000 per year in backbone support from NSF along with other assistance.[7]

Notwithstanding the broadly recognized federal role, the Internet has benefited already from substantial private investments, often through partnerships aimed at extending the infrastructure and experimenting with new services and modes of delivery. For example, IBM and MCI contributed millions of dollars to the NSFNET backbone through ANS, in which they have had equity interests.[8] The Gigabit Testbed program funded by NSF and the Advanced Research Projects Agency (ARPA), credited with accelerating the industrially driven development and deployment of asynchronous transfer mode and SONET technologies, involves substantial commitments of money, time, and talent from telecommunications companies. The launch of certain state and regional

Box 5.1 Federal Funding for the NREN Program

Total federal funding for the NREN program was $114.4 million in FY93 and is expected to be $142.1 million in FY94, with significant amounts from this funding expected to cover the extra costs associated with making the transition from the larger NSFNET to commercial service and a separate research network (the vBNS). The budget request for FY95 is $176.8 million. These figures cover all aspects of the NREN program, including research, application development, and network operation among the National Science Foundation (NSF), the Advanced Research Projects Agency (ARPA), the U.S. Department of Energy, and the National Aeronautics and Space Administration, which account for the largest segments of the NREN budget, and other High-Performance Computing and Communications (HPCC) agencies (the National Institutes of Health, National Security Agency, National Oceanic and Atmospheric Administration, Environmental Protection Agency, National Institute of Standards and Technology, and Department of Education). The ARPA and NSF receive the largest shares of the NREN program budget; the ARPA portion of this program was $43.6 million in FY93 and $48.7 million in FY94, and the NSF portion was $40.5 million in FY93 and $47.9 million in FY94. The new Information Infrastructure Technology and Applications (IITA) component of HPCC, intended to foster government-industry-university collaboration in the development of "technologies needed to improve effective use of the NII," was budgeted at $156.0 million in FY94, with ARPA slated for the largest segment, $95.6 million. NSF has the next-largest requested segment, at $19 million. The FY95 HPCC budget request emphasizes the IITA component in the proposed increments with an 81 percent increase, resulting in a total IITA request of $281.6 million.*

---

*Committee on Physical, Mathematical, and Engineering Sciences, Federal Coordinating Council for Science, Engineering and Technology, Office of Science and Technology Policy. 1994. *High Performance Computing and Communications: Toward a National Information Infrastructure.* Office of Science and Technology Policy, Washington, D.C.; and Committee on Information and Communication, National Science and Technology Council. 1994. *High Performance Computing and Communications: Technology for the National Information Infrastructure.* Supplement to the President's Fiscal Year 1995 Budget. Office of Science and Technology Policy, Washington, D.C.

networks, such as NYSERNet, and confederations of some, such as FARNET, also drew on substantial private investment. *Thus, private-public partnerships have been essential to Internet growth for some time—the federal Internet investment has always been bounded and leveraged.* The private component will inevitably rise with investments in network infrastructure and investments in the commercial provision of information resources.

Despite the exponential growth of the Internet, commercial networks that provide Internet service have grown more slowly than some believe they might have. The withdrawal of NSFNET backbone support was motivated by arguments that such federal funding constituted a market-distorting subsidy, inhibiting the entry of competitive providers.[9] Nevertheless there are today at least three-dozen commercial providers of Internet interconnection service. These range from Internet-access specialists, such as AlterNet, ANS CO+RE, DELPHI, PANIX, and PSI; to local or regional networks, such as Westnet, BARRnet, CERFnet, LosNettos, NEARnet, NYSERNet, Sesquinet, and SURAnet; to full-service telecommunications carriers, such as AT&T, MCI, and Sprint. Entry into the Internet interconnection market indicates that commercial service without subsidy appears viable even if, as is often the case in new markets, it takes time for service to be profitable.

Inferences from Internet-related business successes should be drawn with caution. Successful delivery of Internet service to the research and education communities is not necessarily an indicator that the same service will succeed commercially, because the research and education communities are not a microcosm of society. They are unusual in that their use of the Internet has been subsidized, their priorities and values regarding the kinds and conditions of network use are nonproprietary (see Chapter 4), and most Internet users are served en masse from university or industrial campuses. Competition among Internet access providers for a budget-constrained clientele suggests that there are negligible margins to support donations of services, cross-subsidies, or hypothetical new taxes or other mechanisms aimed at generating support for needy users.

The planned private-sector, multibillion-dollar investments in enabling infrastructure suggest that it is easier for businesses and investors to see profit in home entertainment than in the information-sharing activities on the Internet. This situation is notable given the minimal experience with the new entertainment services touted for the information infrastructure.[10] Indeed, financial analyses in the wake of the aborted Bell Atlantic-TCI merger suggest that investors may become more conservative about profits and opportunities relating to the cable industry (and even telephone companies), potentially constraining the flow of capital.[11]

COST OF NETWORK INFRASTRUCTURE

The cost of network infrastructure is fundamental to both investment requirements and pricing options. Key cost elements include circuits, switching and other internal systems, user systems (computer and communications equipment on customer premises, and associated goods and services, including training), and various high-level and application services. The relationship between cost and price is relatively well established in telephony given its long history and the influence of regulation—but that relationship is poorly understood by many, and often seemingly indirect. The relationship is even murkier in the Internet, which is free of some of the extraordinary requirements found in telephony,[12] and at best a subject of speculation in the emerging NII. Nevertheless, some examination of costs is essential to understand how different pricing options relate to economic efficiency, profitability and investment prospects, and the ability of the federal government to leverage private action in support of policy objectives.

The massive programs of infrastructure development that are beginning to take shape, motivated in part by the possibility of interactive television and video on demand, could bring the benefits of Internet or Internet-like services to a much larger community.[13] The achievement of these benefits is not guaranteed, given that many choices must be made by multiple parties concerning architecture and deployment (see Chapter 2) and that the various enterprises might face higher costs in the short term to provide the kind of open access envisioned in Chapter 2 as desirable for the long term.

Most costs of the Internet are fixed costs (that is, they are independent of usage). Fixed costs for long-distance Internet communication are dominated by the costs of transmission lines (relatively expensive) and routers (relatively cheap within NSFNET).[14] The routers, which interconnect point-to-point circuits obtained from telecommunications carriers, perform packet switching.

The cost of network service to any individual can be divided between the cost of network access and the cost of long-haul communication. Today's network access circuits are either dedicated transmission lines or are shared. Because the sharing that is achieved through packet switching is not common in access networks today, the costs of access are higher than they might be. By contrast, long-distance communication within the Internet uses packet switching to achieve transmission efficiencies that are not possible in the networks used for telephony (Box 5.2).

Today, long-haul circuits operate at 45 Mbps, which costs, at tariff rates, about $45 per mile per month. Very approximately the total number of T3 circuit miles in the NSFNET backbone is about 16,000 miles and so might cost about $720,000 per month.[15] Network operations increase this amount while academic discounts and donations reduce it. This order of magnitude suggests that the cost of long-haul communication for the several million Americans who use the Internet is on the order of

Box 5.2 The Importance of Sharing

A network is a shared resource. It is composed of expensive resources, such as high-speed long-distance trunks, which are too costly for even the largest users to purchase exclusively. Instead, the network is organized to share those resources among all the users in a cost-effective and productive manner. Different networks have different approaches to sharing, but the concept is always present.

The idea of sharing expensive resources is a very familiar one—most people do not own their own airplane, hospital emergency room, or highway system. The need is occasional, and sharing is a reasonable approach. However, people also understand that this sharing can lead to limitations in access to the service: there is not always a plane going exactly where you want at the time you want to leave. And even if there is a suitable plane, there may be excess demand, in which case some potential users may be deferred or delayed. Queues for service form even at emergency rooms.

In the current U.S. telephone system, there is almost always enough capacity to serve all users, but under unusual circumstances, such as Mother's Day or a natural disaster, the system can become overloaded, and some calls cannot be placed. Most consumers, even if they find this frustrating, understand that it is a reasonable compromise that produces lower overall costs.

The technical term for overload is "congestion." Congestion and its control have been a major topic of study in the network research community, since proper utilization of network resources is key to cost-effective operation.

Most Network Costs Are Fixed

From the perspective of the network provider who actually constructs and installs the transmission facilities, most costs are fixed—related not to the instantaneous level of usage but to long-term traffic statistics.* Rights-of-way must be obtained, workers must be paid to install and maintain facilities, and so on. All these costs are unchanged whether the links are loaded or idle.

This situation implies that the network provider must manage the usage of its expensive resources, because both congestion and underutilization are undesirable. Congestion leads to user dissatisfaction, and underutilization leads to reduced revenues. So long as the network remains uncongested, the greater the number of users it supports (or, alternatively the more traffic each user offers), the greater the potential for revenue and for cost-recovery and profit.

Again, the comparison with the airline industry is informative. Airlines must pay for planes and operating costs whether they fly full or empty. So they employ a variety of schemes to sell every seat. Occasionally, they overbook a flight (which is the equivalent of congestion), and to deal with that problem they offer rewards to passengers who will voluntarily defer their travel to a less crowded time.

Network Traffic Is Bursty

In digital packet networks such as the Internet, resources are allocated to users on a millisecond basis, whenever the computer has a burst of data to send.

Measurements show that network traffic generated by computers is extremely "bursty." (That is, most of the time a computer has nothing to send, but every so often, the computer transmits a short burst of data.) This makes sharing a high-speed transmission link among many users very desirable, since each burst from a user can be sent at the link speed, so that it reaches its destination with the least delay. Not only is the offered load bursty, but it also comes in all sizes. A user doing a remote login to a host is probably generating a peak data rate of a few hundred bits per second. A video stream might be 10,000 times bigger, at a few megabits per second. And a fast bulk data transfer between today's high-performance workstations might be 100 times faster than that, at a few hundred megabits per second peak rate. This lack of a suitable model of data traffic makes the analysis of the statistics of sharing rather complex.

---

*The number of trunks installed and the number of operators employed at any time are determined by a planning process that itself is based on measured traffic.

a few dollars per person per year. However, the bigger cost barrier lies in the access circuits, rather than in the long-haul portions of the network.

The committee's vision of an NII characterized by an Open Data Network (ODN) architecture and including both entertainment-telephone-cable TV and general data communications services implies a need for two-way service that is comparable in speed to that needed for compressed video—on the order of 2 to 4 Mbps, based on current industry discussions.[16] At present most people manage with much more modest speeds, such as those accommodated with dial-up modems (typically 9.6 or 14.4 kbps) over voice-grade lines, because these are compatible with existing telephone plant; those homes with integrated services digital network access can communicate at 128 kbps (or twice that, after compression),[17] and reasonably priced Ethernet service over cable has been introduced in portions of California and Massachusetts. But as experience with the Internet demonstrates, networked file systems quickly lead to lengthy data transfers and the need for higher transfer rates. In the near future interactive graphics and multimedia documents will be a common means of information exchange that demand high speed. Accordingly, the trend on university campuses is toward Internet access at 1.5 Mbps and 45 Mbps. Pressure for these data rates to and from people's homes will emerge (Box 5.3).

The scale of the NII challenge is a principal reason that the federal government has chosen not to contract for a separate network but to transfer the burden of network construction and operation to commercial interests.[18] The research and education communities illustrate the scalability problem even though they are only a part of it: these communities are widely distributed across the United States; they occupy 110,000 K-12 school buildings and over 3,000 higher-education campuses; and the research and educational infrastructure for those who have already left school involves some 15,500 public libraries. These facilities house approximately 47 million school children, 15 million higher-education students, and 3 million teachers and faculty.[19] Moreover, most of these same individuals also work at home—children and older students do homework, and faculty grade work and prepare for the next day of teaching. Serving a widely dispersed community reduces the opportunities for sharing. If only a few sites in a community are connected to a network (e.g., only the schools and libraries), it will be necessary to allocate to these few users not only the cost of the local access link from each site, but also the cost of the link from the community to a more distant network access site. This argues that there will be considerable economies of scale in serving education and libraries through a larger NII whose economic justification comes from other commercial users. *It would be prohibitively expensive to provide high-speed data communication service to a large number of widely dispersed homes and campuses for research and education purposes unless there could be some economy through sharing with other users of a nationwide data communications infrastructure.*

To these quantitative concerns must be added qualitative ones: the priorities, timing, and location for adding new facilities and capabilities to the commercial networks may not be directly aligned with the preferences or needs of the research and education communities—indeed, this differential was a driver of the NREN program.[20] These realities suggest a risk that some members of the research and education community will, in the short or long run, not be served as well by commercial services as they are by the Internet today. The costs of enabling upgrades to the local loop are a key barrier to the rapid and widespread deployment of NII services. See Box 5.3.

Clearly, service providers must cover their costs of capital and operation, which implies some system for charging network users. The price of access to a commercial Internet service today (like the price of access to a regional network) depends greatly on the distance to the nearest net-

Box 5.3 Access Links and Where the Sharing Occurs

Access links (the "last-mile" connections to customer sites) represent an important aspect of system costs. For access links to small sites such as residences, there are few opportunities for sharing the cost among several sources of traffic. This creates a tension between two objectives. If the access link is of large capacity, which is preferable (albeit costly) to permit bursts of traffic into the network, this raises the possibility that the traffic source could produce not bursts but a continuous stream of traffic at this high speed. Providers are thus motivated to impose a high charge for such a link. Alternatively, the access link can be engineered at low speeds, which protects the provider but limits the traffic source. One solution to this dilemma is to engineer a new form of access link that permits high burst rates, but that either prevents or charges for high continuous traffic flows.*

Large institutions (for example universities or corporate sites) have an easier time with the economics of sharing. With a site housing a large number of users, the very bursty traffic flows from these many sources become combined into a somewhat smoother aggregate stream across the access link into the external network. The institution can thus justify purchase of a high-speed link and allocate the cost across the total pool of users.

---

*Note that loop transmission architectures are complex and may require, for cost and/or efficiency reasons, electronic and/or optical multiplexing that is accomplished via equipment deployed within the loop (so that the fiber is no longer dark). Also, hybrid fiber-copper loop architectures may be feasible.

work node. That may be tens or hundreds of miles because there are relatively few points of network connection. Geography is a major reason for cost differences experienced by the University of Colorado and the Massachusetts Institute of Technology (MIT). With the introduction of commercial Internet service, MIT, in Cambridge, Massachusetts, anticipates that its costs may increase on the order of $10,000 per year (assuming both a network point of presence in Boston and a bandwidth-driven access charging scheme), whereas the University of Colorado, in a more rural setting with greater distance to the nearest network node, estimates that its annual costs could rise by as much as $100,000.[21] Averaged across all connections to commercial Internet services today, the average charge for access at 56 kbps is $15,000 per year and the average charge for 1.5-Mbps access is $36,000 per year. As access to high-speed services widens (scale again), the nationwide average cost for commercial Internet access at 56 kbps might drop from today's price of $1,250 per month to a figure that is more like $50 per month.

Today's cost estimates should not be projected far into the future, since new technologies (such as the introduction of more fiber-optic facilities and such digital transport services as ATM) will be cheaper than digital services over voice networks. Both network- and information-related costs are likely to change, for such reasons as increasing competition, deregulation, changes in enabling technology, and increases in scale. Emerging plans for reconstructing the cable TV and telephone access networks will enable at least 1.5-Mbps access for each video channel; they open the door for interactive services.[22] Reconstructed access networks for cable TV, for example, should be able to support Internet access and Internet-like services at much lower cost than is evident today.[23] Efforts are already being made to exploit such connections using special hardware (such as new kinds of modems supporting Internet access over cable systems) and services.[24]

Although the above discussion has focused on line costs, another cost element, under the current Internet structure, is associated with the physical and business aspects of the interconnection of regional, commercial Internet access, and other networks. At present, interconnection is provided by the Federal Internet Exchange (serving federal backbone networks) and the Commercial Internet Exchange (serving commercial Internet access providers) locations. The recently announced interregional CoREN is intended to serve regional networks, and the new NSF plan calls for network access points (NAPs) to interconnect various networks with the anticipated vBNS and with each other. Interconnection entails associated equipment and labor, which create costs. In the telephony environment it has also involved other costs (e.g., settlement payments covering traffic transiting the interconnection), some associated with regulation. Added cost elements in the telephony environment contribute to the observed differential in cost between sending a fax via Internet versus via the telephone network. See Box 5.4.

As the shift to commercial service unfolds, there is no reliable information on how prices will change and what impact those changes will have on patterns of network use. To understand and plan for both the Internet and the NII, it would be useful to have a comprehensive study of Internet service economics, how the cost of that service can be expected to evolve in the coming years, and how it will be affected by continuing growth of the user population, the shift to multimedia communication, and the restructuring that is taking place in the telecommunications industry.[25] Some preliminary efforts at relevant analyses have begun, but comprehensive and authoritative analysis would help to dispel myths and assess the cost-effectiveness of potential courses of action.

Box 5.4 Interconnect Charges in the Sharing Paradigm

Issues of sharing and congestion also arise inside the network. Because the network is not implemented as one monolithic entity but is made of parts implemented by different operators, each of these entities must be separately concerned with achieving good loading of its links, avoiding congestion, and making a profit. The issue of sharing and congestion arises particularly at the point of connection between providers. At this point, the offered traffic represents the aggregation of many individual users, and thus cost-effective sharing of the link can be assumed. However, if congestion occurs, one must eventually push back on original sources. Options include pricing and technical controls on congestion; there may be other mechanisms for shaping consumer behavior as well.

Since each of the network providers carrying the user's traffic must recover costs, the user payments must be shared among all the relevant network providers. One way for this to be accomplished is for each provider who carries traffic for a user to bill that user separately. However, this approach presents a serious technical challenge for connectionless networks, such as those that use the Internet protocols, since the network protocols of today (such as the Internet protocols) do not have the necessary information in the data packets to permit trustworthy accounting, and accounting at the packet level may require mechanisms of considerable cost and complexity.

The alternative in use today among the interconnected providers is flat-rate payments. This approach is consistent with the observation that most costs are independent of level of usage, and it provides to both parties the assurance of known payment levels.

PAYING THE PRICE

Whereas the committee's vision of an open NII suggests ready access to the network by individuals wherever they may be, beginning with their homes and places of work, the goals of research and educational networking have centered on access by an institution (e.g., a school or a laboratory), which in turn provides access to individuals within the institution. Having recognized this, the NSF Connections program (and even the NTIA Telecommunications and Information Infrastructure Assistance Program and other elements of the administration's NII proposals) focuses on achieving some kind of an institutional connection as a first step; similarly, NSF's Stephen Wolff has spoken[26] of a three-pronged strategy for community access: public-access terminals in a school, a public library, and a county extension office.

The apparent institutional focus arises not because it is sufficient, but because it is more affordable and practical than attempting immediately to connect all relevant individuals within publicly supported education, research, library, and health-related institutions.[27] Also, institutions typically provide the needed expertise for support; a substantial but often unrecognized cost relates to training and ongoing support for users.[28]

However, it is important to recognize that institutional access presents a different concept of national scale. It also is associated with certain kinds of gaps in support, such as the lone researcher at a small college who might not be able to get access (whereas new investigators or investigators in new fields at connected institutions might). That is, all institutions are not alike: some institutions are able to afford good access and some cannot afford any.[29] Moreover, those with difficulty affording access, per se, have difficulty affording those complementary goods and services that allow them to realize the benefits of access. These range from adequate local infrastructure, to user training and support, and to access to certain information resources and services available over the network. The Pacific Bell proposal to provide connectivity for public schools and libraries was notable for its provisions to "field dedicated resource teams" to assist in the use of the capabilities being provided as well as to collaborate with universities and colleges to enhance the treatment of educational technology in the teaching curriculum.

Today, most individual users of the Internet gain access through the institutions that employ or educate them and do not see the costs or charges associated with their individual use of the network. Institutions for research and education do pay for backbone access: they typically pay access charges, often through a regional network, and they have had free use of the backbone network, which supports long-distance communication. By contrast, commercial users of the Internet (e.g., corporations) pay both for access and long-distance communication (since the NSF's acceptable use policy has restricted use of the NSFNET backbone).

Many institutions, such as high schools and public libraries, have only indirect access to the Internet (through a time-shared machine attached to the Internet and operated by a service provider, usually accessed at low speed, e.g., 9.6 kbps or less). These users typically pay for dial-up access (telephone service) and pay a fee that is a function of the time spent using the network.[30] Depending on location and usage, such connections can be very expensive. For example, a pilot project for connecting rural libraries in New York state to the Internet found that several participating libraries were paying $175 to $350 a month in long-distance telephone charges because there was no point of presence in their local area, a figure that could increase with usage.[31]

Economic analysis argues that individuals must encounter accurate costs in order to make appropriate choices or trade-offs among alternatives and in order to be as effective as possible in their use of whatever level of funding is available. If alternatives appear to be free, individuals will choose those that appear the best to them, regardless of cost; a free alternative is always consumed to excess. If *A* and *B* are both free, *A* and *B* are consumed to excess. If *A* is free and *B* has an appropriate cost-based charge, *A* is overconsumed and *B* is either underconsumed or consumed at a proper level. Thus, it can be hard to assess what choices would otherwise be made in an environment where cost or price information is distorted. Among the choices to be made within the research and education communities are efficient trade-offs among the various inputs used in producing research, education, or library services, including those that complement the use of networks (e.g., desktop systems and other resources for gaining access) and those that substitute for networking (e.g., other means of communication and gaining access to information).

The provision of access to the Internet without recouping full costs from users (institutional or individual) laid a foundation for two phenomena. One is the extraordinary growth in networking and applications that excites people about the Internet. The other is the growth in a sense of entitlement to low-cost (even no-cost) access within the research and education communities. These outcomes stem from the focus on the Internet as a research-related enterprise; "free" access to ARPANET and its successors was granted to the institutional communities not only to facilitate the conduct of research, in part by broadening access to such shared resources as federally funded supercomputer centers, but also because of the research value of seeing how use would develop. This access subsidy was apparently provided without making compensating changes either in the overall levels of institutional support or in the allocations of that funding among inputs for which network access and network resources were substitutes or complements. This situation implies both that there was a net increase in the effective level of funding to the institutional community and that allocation decisions among inputs have been distorted. For example, more network resources and their complements are being used than would be used (even at the same effective overall funding levels) if direct payments for such resources were required.

Thus, the Internet experience with respect to cost supports conflicting interpretations: as an experiment, it suggests a potential for substantial increases in the amount of networking, which is believed to generate increases in the quality and value of network-based activity, but the basis for that expectation lies in what economists might view as excessive use of networking to date, inasmuch as pricing to users was below cost. However, as in other network settings, the Internet experience also points to the benefit of initial subsidies to jump-start use in situations where the value of consuming a product (in this case, networking) increases as a function of the participation of other consumers.

Imminent Short-term Increases

With the move to commercial service, institutions are expected to absorb the costs previously covered by the NSFNET backbone payments and those covered by a declining schedule of support payments for the regional networks. Together, these NSF payments amount to about $30 million;[32] divided among connected universities, the increase in cost is expected to average a few thousands of dollars. Although absorbing this increase may cause some hardship in the short term due to transitional budget-juggling[33] or for institutions for which service is inherently costly (e.g., those far from network access points), the order of magnitude is sufficiently low to suggest that there will be little hardship on average for institutions already connected to the Internet. In addition, NSF has discussed recouping funds from the scheduled reductions in support for regional networks and making them available on a competitive basis to institutions seeking to effect high-bandwidth connections. Thus, the change in NSFNET support is causing more apprehension than circumstances appear to warrant.

RECOMMENDATION: Transitional Support

The committee recommends that temporary subsidies of education and research institutions be considered in cases where the commercialization of the Internet generates exceptional funding distortions.

The last decade has seen a powerful transformation of the activities of higher education and research in those segments where networking has become integral. Side effects of the shift away from a federally funded NSFNET backbone may include temporary disruption and financial hardship for some members of these communities, although overall the impact is expected to be limited.

Costs of Local Infrastructure and Access to Services

There are two other elements of concern to Internet users: local infrastructure, and the costs of accessing services once network access is achieved. Although financial support for the NSFNET backbone and the regional networks constitutes a significant amount of money, it is dwarfed by other elements of the cost of using the Internet. Wide area network service typically does not include installation, operation, and maintenance of the local infrastructure on customer premises. Local area networks (LANs), workstations and other desktop equipment, communications and interface software, and the staff that manage or support these systems cost far more than wide area network access and use.[34] For example, one industry survey suggested that the typical corporation's average five-year cost of owning a personal computer is $40,000, reflecting substantial installation and learning-related costs—which have risen over time with the complexity of the technology.[35] This situation is recognized in research as part of a larger, chronic problem in attaining and upgrading infrastructure (facilities, hardware, software, and support). Local infrastructure elements must usually be paid for out of funds that are distinct from those that pay for wide area communication obtained from a commercial provider.

The cost of the local infrastructure is of particular concern to people building up the capability to use the Internet for the first time. For example, it is particularly acute for the K-12 education community, which has lacked adequate electrical power and telephone access, let alone access to computing and communication-related equipment, software, and technical support. A pilot project to connect rural public libraries to the Internet involved an investment of about $3,000 for a single basic package of hardware and software per site.[36] The absence of targeted support for local infrastructure means that observed spending reflects trade-offs and choices made among a wide set of alternatives, given the available budgets. However, those trade-offs and choices may be uninformed when decision makers have not experienced network access. Experimentation, yielding insights into both the potential benefits and the costs of information infrastructure, may help people to learn how to lower costs.

Local infrastructure is essential because many of the most visible networking services involve capabilities provided outside the backbone network and often go right to the end user's computer.[37] Capabilities such as multimedia electronic mail and distributed database retrieval place a very heavy reliance on the end user's systems; the networks are relegated to pathways for the flow of bits, while much of the action (and investment) takes place outside the network. Thus, a critical infrastructure challenge and cost is in developing the intrasite infrastructure. In California, for example, that challenge pertains to 136 higher-education campuses and approximately 7,500 school sites and 800 libraries; a rough estimate is that the cost will exceed $2 billion using today's technologies and costs. Assuming these costs can be met, the availability of small-scale, personal computer-communications systems is what enables individuals to introduce new information resources and services, one of the major benefits associated with the Internet and the NREN program.

It is important to note that the cost of network access is only a piece of the cost of information infrastructure access—there are additional costs associated with access to and use of information resources, costs that may be separate from or bundled in with network access. Database vendors, for example, provide access to their offerings at prices that may include a monthly subscription fee plus connect time and document or page charges or, increasingly, may be priced on a flat-fee (per month or other subscription-period) basis.[38] Because of the economies of scale in the production of information, varying prices charged to different classes of users—including charging lower prices for some who otherwise could not purchase the information—may be desirable for optimizing the production of information products.[39] Although the price charged reflects many considerations (ranging from ownership of intellectual property to the costs of assembling and packaging the information for customers), the actual costs of getting into and extracting information from a database are very, very low. Most of the marginal costs for an individual user are associated with physical access (including, in addition to communications links, the software that adds a user's address to the network for billing and connecting purposes).

Not all information resources require additional payments. In the Internet environment, there are many information resources offered without charge by individuals, government agencies, and a variety of organizations—data, research results, "shareware," "freeware," certain government information (increasingly becoming available over the Internet), and so on. Even in the commercial environment, advertiser-supported services allow for some resources to be available without additional charges.

It is easier to gauge changes at the institutional than the individual level, although both will affect the level and kinds of use of networking in the research and education communities. Within institutions, questions remain about how increases in costs will be absorbed, what administrative and management changes will be made in response to an increase in costs, and the extent to which individuals who are disproportionately large users of network services may confront sudden large bills for their use. In this area, the central questions are, (1) Should individual users face charges based on their level of use? and, (2) Is there a preferred approach to pricing for infrastructure access and use? The second question, in turn, has two parts: (1) How should individual use charges be covered? and, (2) Is there a preferred mechanism for delivering a subsidy, if any?

Usage-based Pricing

Given the "bursty" nature of data communications traffic (see Box 5.2) and the wide variation in demand for capacity over time and in terms of bandwidth, *any network that attempts to realize effective sharing must include sophisticated mechanisms to shape user behavior.* The rate at which traffic is offered must be regulated to generate reasonable link loadings without serious congestion. These controls can take several forms, including pricing and technical mechanisms.[40] Another argument for user charging is the expectation that, with finite resources, achieving a truly national information infrastructure will require as broad a base for cost recovery as possible. If the NII is really going to benefit a wide range of people, revenues should be derived at least partially from the individuals using the services. The High-Performance Computing Act of 1991 contemplated that the NREN could "have accounting mechanisms which allow users or groups of users to be charged . . . where appropriate and technically feasible, for their usage of the Network."[41] See Box 5.5 for a list of key cost elements in the Internet context.

As noted above, economic analysis argues for usage-sensitive charging where costs are usage-related. Such charges both enable users to choose appropriately among alternative purchases and provide a check on excessive network use. If unlimited network use is available for a flat fee, there may be little incentive for restraint; in the network context, the result is congestion, as on an overused freeway, which imposes costs on other users. Specifically, those who overuse the network may reap all of the benefit but suffer only a fraction of the cost imposed on the network in the form of congestion.[42] Congestion has not been a major problem on the Internet recently, both because of increases in capacity intended to accommodate growth (and overcome past congestion problems) and because most applications have been tolerant of the congestion control mechanisms in the network, which delay traffic in response to heavy demand or load. Since most Internet costs are relatively independent of level of use, in the absence of congestion the incremental cost of extra traffic is essentially zero.[43]

Internet congestion appears to depend on the time of use relative to periods of peak load. This situation is similar to other network contexts, where, for example, variation in pricing for peak and off-peak use can help to balance network load and accommodate users with fewer resources. However, some applications (such as nontime-sensitive file transfers) are more suitable for off-peak timing than others (e.g., real-time communication, some e-mail).

The Internet has evolved, perhaps by accident, with a payment model based primarily on a fixed price for access and a service model that

Box 5.5 Matching Prices to Costs

In general, the prices that users face should reflect the resource costs that they generate so that they can make intelligent decisions about utilization of resources. In the case of the Internet, there are several costs that might be considered:

- *The incremental costs of sending extra packets.* If the network is not congested, this is essentially zero.
- *The social costs of delaying other users' packets when the network is congested.* This is not directly a resource cost but should certainly be considered part of the social cost of a packet. Users bear this cost through delay and dropped packets and would often be willing to pay to reduce congestion.
- *The fixed costs of providing the network infrastructure.* As has been observed, this is basically the rent for the line, the cost of the routers, and the salary for the support staff.
- *The incremental costs of connecting to the network.* Each new user (or network of users) connection to the Internet involves costs for access lines and switching equipment.
- *The cost of expanding capacity of the network.* This will normally consist of adding new routers, new lines, and new staff.

SOURCE: MacKie-Mason, Jeffrey K., and Hal R. Varian. 1993. "Pricing the Internet," prepared for a conference, Public Access to the Internet, John F. Kennedy School of Government, Harvard University, May 26-27, 1993. Prepared April 1993; November 26, 1993, version.

shares the available resources equally among all those who at one moment want to use the network. In times of low traffic, the amount of service that any person gets under this regime is not determined by the funds that he or she has available but by the capacity that exists to serve that user. (See discussion on best-effort versus guaranteed bandwidth service in Chapter 2.) By contrast, the telephone network charges either according to the amount of use that is made of the network, or at rates independent of usage (at least up to some volume); no matter how many users there are, it offers a constant quality of service. The Internet offers variable-quality service (and uses technical means to control congestion) for a fixed price, and the telephone network offers a fixed quality of service priced according to the amount used.

This distinction will become less clear-cut and less sustainable as workstations are equipped to support audio and video communication. Internet users will probably wish for predictable quality of service as an option when they begin to use their computers as substitutes for telephones and videophones and as audio, video, and multimedia applications proliferate. Moreover, it would be more efficient to assign higher priority to scarce bandwidth resources for services with a higher value (such as real-time medical imagery versus leisure-time game images).[44] Both schemes have their merits and should be available in the Internet after the switch to commercial service; the implication is that there will be correspondingly different pricing schemes for different kinds of use (e.g., very low prices for text e-mail and higher prices for real-time video).[45]

It is useful to point out that best-effort service and flat-fee service are not synonymous; it is possible to have usage-based best-effort and flat-fee guaranteed service arrangements. Note also that any kind of traffic- or usage-sensitive pricing scheme presumes that detailed accounting information is available to support it. Pricing structures and their relationships to different models for delivering services constitute an area under discussion within the Internet community.[46]

Flat-fee Pricing

Most research and education institutions are highly resource-constrained: they depend primarily on public financing, they must hold to fixed annual budgets, and, like many in industry, they have difficulty with costs that are not predictable over the course of a year. Research libraries' experiences with direct charging for commercially supplied information services—which libraries consider very expensive—and recognition by some researchers that they generate substantial volumes of traffic lead to deep concerns about the volume and incidence of direct user charges. Consequently, research and education institutions would prefer to obtain networking and information services for a flat annual fee rather than be charged directly and variably for the traffic that they generate.

Commercial providers of Internet access, like providers of telecommunications services generally, offer a range of charging schemes; variation in pricing is one dimension along which providers compete.[47] There are flat-rate schemes that vary according to the speed of the access line (which is presumed to correlate with volume of communications traffic), and usage-sensitive schemes that are based on the number of messages transmitted (or other transactions), the time spent using the network, or the number of packets transmitted. There is a trend toward flat-rate schemes, in response to emerging commercial customer demand and complaints that usage-sensitive charging is too expensive and inhibits use.[48]

Different payment structures share risk between the user and the provider in different ways. A usage-based structure has risk for the user, in that unexpected usage (e.g., from misunderstanding or poor planning) can generate large expenses, which may not have been budgeted. At the same time, usage fees have risk for the provider, in that if the users curtail usage to control costs, the revenue to the provider will drop. Flat fees or access-based rates have the mutual advantage that the cost to the user and the revenues to the provider are known in advance. In exchange, flat fees offer the risk of congestion, if the users are not prevented from offering excess traffic. With flat-rate schemes, the provider assumes more risk (associated with variation in usage) than with usage-based schemes, but the impact depends on how much cost varies with use—and for much use there is little such variation. Also, low-volume users effectively subsidize high-volume users when both groups pay the same fee.

One issue as these developments unfold will be how costs vary by kind of traffic. Technically, a bit is a bit, as is observed in Chapter 2, but voice and video bits are priced very differently today. Voice-level charges discourage video traffic (which requires far more bits), whereas video-level pricing would allow voice to be carried essentially for free. These differentials contribute to carrier concerns about proposals to unbundle the basic transmission facilities from some form of basic service.

Given observed market conditions, the committee assumes that a variety of pricing schemes will be available to consumers of Internet service.[49] It recognizes that flexibility in the way that Internet service is priced would allow those that make light use of the network to avoid the extra cost implicit in a flat-rate price that is also designed to accommodate heavy users. Flexible pricing also could permit a situation in which those with fewer resources receive lower-quality service or service with fewer guarantees as to delay or bandwidth, while those with more might achieve more quality or service guarantees; with flexible pricing there might be a series of service packages offered at different prices.

A set of service packages with different flat fees and corresponding different levels of service covered by the fee would be desirable. This would allow users to be differentiated according to need, so that a person who wants to send video will pay more than a person who sends only electronic (text) mail. The concept is consistent with the two-tier service recommendation of the 1988 Computer Science and Telecommunications Board (CSTB) report, *Toward a National Research Network*;[50] a proliferation of users, capabilities, and services (both current and anticipated) suggest a need for several pricing packages, not just two.

Network providers can use pricing and other means to balance load and to shift usage to off-peak periods.[51] In addition to congestion-related variation in pricing based on time of use, suggested above, another timing issue relates to making long-term versus spontaneous arrangements to use above-average amounts of network capacity. Users who anticipate only occasional demand for very high speed service can be encouraged to make an explicit reservation in advance, as an alternative to paying for the option of sending that traffic spontaneously. It is necessary to determine how to handle a situation in which someone generally has low-level needs but, for example, once or twice a year wants to do a video conference.[52]

A flat fee for network access offers other advantages relating to the larger goal of developing the NII. Experience with data networks is still developing, and people are uncertain about how much network capacity is consumed when a particular program or application is executed. Flat-fee charging removes the tension that is otherwise present in these circumstances, and it frees the user to experiment with new capabilities without fear of bankruptcy. Moreover, as noted above, although costs of new applications involving audio and video will be relatively usage-sensitive, dominant applications today (e-mail, file transfer) typically are not.

In contemplating pricing options, there is an assumption that growth in the volume of users will result in lower average prices, given that there are economies of scale. Another assumption is that an increasing number of service providers creates competition that will also place a check on prices. Further, pricing schemes may motivate new protocol developments that can result in lower costs. All may take time to unfold, and neither assumption is inconsistent with some degree of continued government support for research and education users over the near term. As a result, *the ideal pricing structure is likely to change over time.* The argument for flat-fee charging, in the near term, to stimulate experimentation and use, extends beyond the research and education communities to business and the general public. Usage-sensitive pricing might be preferable in the long run, when network-based applications are as mature and familiar as voice telephony, photocopying, postal service, and other information-related transactions for which usage-sensitive charging already exists.[53]

Covering User Charges (Subsidies and Mechanisms)

Although the level of financial support is relatively small, symbolically the withdrawal of NSF support for the NSFNET backbone and eventually the regional networks signals a need for users to acknowledge and cover the costs of their use of network-based resources. In the public financing context of research and education, many fear, this situation may constrain the use of networking and network-based information resources inasmuch as budgeting is zero-sum in the short term. However, other current cost factors may change (for example, changes in state regulations may lower costs to research and educational users), and spending on network-based resources may result in less demand for other resources (such as books, as has been seen in the library environment).

Eventually, use of network-based resources in research and education may be budgeted like use of other communications and information resources. This has already happened in some instances, and it is typical of for-profit enterprises. On the other hand, a number of mechanisms are available to support networking and access to networked information resources should such expenditures be targeted as a matter of policy—which would be consistent with the overall policy decision to promote an NII. The drawback to subsidizing infrastructure-related expenses specifically is that there are many information products involved in education and research, and subsidizing one subset is done at the expense of others that may be at least as useful.

Many academics receive federally funded research grants awarded competitively based on proposals. A grant proposal typically includes line items for cost components such as travel, telephone use, photocopying, and computing. There is also provision for institutional overhead. In the future, network or information infrastructure access will probably show up either as an additional line item or as a component of the institutional overhead. As it is, today mission agency (e.g., DOE, NASA) funding is generally not directed to network users as such (except in the case of backbone support), but to research principal investigators who must decide on the prioritization of network needs versus other programmatic needs, given some constrained funding level.[54] *The advantage of including network and information access along with other cost items is that it allows the individual researcher to choose how best to allocate resources and to make choices among networks, telephones, and other alternatives.*

Practical arguments are often mounted against a grant-based approach. For example, if networking is treated as a line item, there is some concern that the resources available to the recipient will be reduced to cover institutional overhead costs, as happens with other kinds of expenditures. Another concern with individual project recovery is the administrative burden of tracking yet another cost item. These are among the arguments that were advanced in opposition to the suggestion made in the 1988 CSTB report to provide network-service vouchers to individual researchers to be used in covering network-related costs. Also, as discussed in "Equity" below, a research-grant vehicle could serve only that portion of the research and education communities that receives such grants.[55]

It is important to distinguish arguments raised against a specific kind of mechanism (e.g., a voucher for an individual researcher) from arguments concerning a generic type of mechanism. As MacKie-Mason and

Varian point out, the concern about Internet access by "poor" users relates to the distribution of wealth, not pricing. Vouchers or lump-sum grants are often recommended by economists as a relatively efficient mechanism for redistributing resources.[56] Vouchers effectively provide a new currency, in this context one useful only for network-related services.

Given the administrative problems in routing funds to individuals, it may be more convenient to target funds to service providers rather than to end users directly. This approach seems relatively straightforward in the case of network service providers, but raises additional implementation considerations if "service provider" is expanded to include information service provider.[57] That convenience has been implicit in federal procurement of network facilities and services through the NREN program to date.

One possibility tied more closely to infrastructure use is a system of vouchers that are directed to service providers—a much smaller group than the end-user population—based on the number of subsidy-eligible users they serve.[58] Such a system would place the administrative burden on the service provider, while allowing users the freedom to choose a service provider and thereby preserving competition among providers for customers. A limitation of a provider-based voucher system is that it covers only a portion of the relevant costs, the costs of the network-based service. Local costs for equipment, software, and support as well as costs for information resources would still have to be covered through conventional budgeting, including grant provisions and other sources of funding.

Deriving Specific Funds

Compounding the distortions that may arise from targeting funds for use of information infrastructure are problems in generating those funds. After considering the pros and cons of specific options, the committee decided against recommending specific options for generating funds that explicitly or implicitly amounted to taxation. For example, taxing communications providers (including increasing existing taxes) could yield revenues to defray expenses incurred by research and education users of the Internet, but at the likely cost of slowing investment, raising prices, and triggering other side effects common to excise taxes.

If there are to be subsidies for research and education users, consistent with the NREN program and also with the connectivity goals advanced for the broader NII, funding out of general revenues would have the benefit of encouraging more explicit balancing of claims for limited federal funds. (The committee recognizes that there are ongoing discussions in government and industry on possibilities for a broad-based pool to support universal access for low-income individuals.) General revenues and tax incentives also allow for a broader spreading of costs commensurate with the broad distribution of benefits anticipated for the NII.

Equity

A central issue in the move toward greater user responsibility for infrastructure expenses is the incidence of costs. (See "Equitable Access" discussion in Chapter 4.) There is concern, under the new arrangements for Internet service, that what users get may depend on how their work is funded. The NREN program illustrates one divide-and-conquer approach: the NSF intends to provide direct support for very high bandwidth scientific research. For this it has commissioned construction of a small network, the vBNS. Other Internet users will receive backbone network service from commercial providers. Clearly, NSF recognizes that some needs are easier to meet "affordably" through the marketplace than others. It implies that as a matter of public policy, some big users will receive substantial direct public support for their networking. The rationale for this arrangement reflects the nature and value of the research program.

To the extent that a funding agency has redirected funds to cover the cost of Internet use, grantees may see no substantial change as a result of the switch to a commercial Internet and user charges. If you do research and your funder thinks that you need high-powered networking, your grant may provide for vBNS access privileges, you may get more resources for networking, you may get access to ESnet or NASA Science Internet. If your state government or institution wants to play in the information age, you may get basic connectivity, perhaps more. But if you don't satisfy a funder's test and if you don't live or work in the right place, you will have to finance access and services yourself.

Individuals in the research and education communities routinely confront differences in financial support, both by field and by institution, and in what is considered affordable. Even telephony is regarded as expensive by some school administrations. High-energy physicists and earth scientists have relatively generous support for networking, and for computing, whereas historians are more likely to have none. A computer scientist in one of the leading university departments would expect to have a leading-edge workstation, whereas a computer scientist in a four-year college may at best have access to a lesser machine, and those studying natural languages may have access to the most basic of personal computers. Within the research community, some disciplines have a relatively healthy level of grant money that could be used to support networking needs, while other disciplines are more resource-constrained. Isolated researchers in rural small colleges face greater difficulty gaining access to networks and using them to collaborate than others better situated. Budgeting for K-12 education, small colleges, and public libraries tends to be even more constrained than for research. Recognition of these disparities fueled the growth and broadening of the NREN program from its roots in supporting supercomputer center access for an elite group of researchers.

The discrepancies among specific disciplines or institutions may be seen as a reflection of higher-level budget decisions about where limited federal funds are best spent, allocations that result from the political process of determining where federal and state funds are spent among research, education, and other alternatives and within the categories of research and education. In that context, there are winners and losers, richer and poorer groups. The difficulty faced by K-12 schools, small colleges, and public libraries in becoming connected flies in the face of a fundamental premise of the U.S. public educational system: there should be equal opportunity for all students—in large schools and small, in rural areas and in the cities, regardless of income. Observed lack of such equality affects access to books, teacher salaries, athletic programs, class sizes, and so on, which compete with Internet access for scarce resources.

Many within the research and education communities believe that the research and education enterprise at large benefits from the broadest possible involvement in information infrastructure. Achieving that broad involvement will require some means of addressing the financing needs of those unable to afford to participate, recognizing that such subsidized participation may be limited in scope or volume. The payoff may be neither direct nor immediate, but it may involve a lowering of costs and federal financing needs over the long term, as suggested by telecommunications economist Roger Noll:

> An educational system that produces a declining fraction of people who think reasonably clearly about decisions under uncertainty and about making inferences from information is fundamentally inconsistent with an information technology sector that offers increasing rewards that are available only to the educated. Of course, federal policies are always in some loose sense competitive substitutes, but in this case the complementarities are especially strong. A public policy that emphasizes enhancing the network and high-end terminal equipment largely for business purposes, but that is not accompanied by a program to enhance public education, and in particular to get computers into the schools and to create data bases and software for use by ordinary citizens, is unbalanced at best. The social value of new information technology is maximized if the maximal number of people can use it, and the act of maximizing the number of informed users also serves to minimize the extent to which redistribution of wealth will motivate the adoption of the technology. Thus better education in how to use information is an essential component of rational public policy toward the information sector.[59]

NOTES

1. According to Roger Noll, "[An] important characteristic of information is that it can be used to make better use of other valuable resources. Information that saves time or leads to better decisions increases the opportunities available to a user, thereby reducing costs in the sense the term is used in economics. Thus, a more expensive computer or a telephone network that can transmit more information per unit of time may reduce costs. But in a large organization, an office head contemplating the purchase of a new computer or telephone system may see only a budget constraint, and not the implicit value of the increased productivity of office employees that these technologies allow. And, a state public utilities commissioner may be held accountable for only the price increase that must accompany an enhancement to the network if the telephone company is to remain solvent, not the reduction in total costs that the enhancement can bring forth" (p. 29). Noll, Roger G. 1993. "The Economics of Information: A User's Guide." Pp. 25-52 in *The Knowledge Economy: The Nature of Information in the 21st Century*. Institute for Information Studies, Nashville, Tenn.

2. Federal support for elements of today's Internet has been reduced in the past, from decommissioning of the ARPANET to diminishing support for Internet engineering and standards-setting activities, a trend that led to the establishment of the independent Internet Society.

3. The Pacific Bell announcement was linked to a state government challenge to link and equip every classroom in California for "full-speed superhighway access by the year 2000." To meet that objective, the Pacific Bell announcement anticipates not only basic data and video connectivity, but also access to video-on-demand and other forms of interactive multimedia for public schools and libraries as the company deploys its interactive broadband network. Assuming the proposal is approved by the California Public Utilities Commission, Pacific Bell would wire targeted locations within each institution, install service (four ISDN lines) for free, and waive usage charges for one year after installation. It also intends to work with the Public Utilities Commission to develop a special educational access rate. "Pacific Bell to Link Public Schools and Libraries to Communications Superhighway," February 14, 1994, press release distributed electronically.

4. White House Office of the Press Secretary. 1994. "Remarks by the President in Satellite Meeting with California Newspaper Publishers," February 12, electronic distribution.

5. As noted in *Toward a National Research Network*, advanced NREN (the original "phase III") technology requires research and innovation. Computer Science and Technology Board (CSTB), National Research Council, 1988. *Toward a National Research Network*. National Academy Press, Washington, D.C. (CSTB became the Computer Science and Telecommunications Board in 1990.)

6. Mandelbaum and Mandelbaum describe how both the regional networks as they are currently organized and the larger three-tier model for NSFNET were "not an a priori construct." Mandelbaum, Richard, and Paulette A. Mandelbaum. 1992. "The Strategic Future of the Mid-Level Networks." Pp. 59-118 in *Building Information Infrastructure*. Harvard University Press, Cambridge, Mass.

7. Briefing to the committee, June 1993, Stephen Wolff, National Science Foundation.

8. The Merit proposal to NSF included provisions for $5 million from the State of Michigan for facilities and personnel, approximately $6 million from MCI in reduced communication charges, and $10 million from IBM in equipment, installation, maintenance, and operation. In September 1990, Merit, IBM, and MCI formed Advanced Networks and Services (ANS) Inc., a nonprofit corporation that was assigned Merit's responsibilities under the Cooperative Agreement with NSF. See Office of Inspector General, National Science Foundation. 1993. "Review of NSFNET," NSF, Washington, D.C., March 23.

9. Some observers, for example, have raised concern about how the NSFNET coopera-

tive agreement was structured and implemented. These concerns were manifested in electronic discussion groups over the Internet and eventually in congressional hearings in the summer of 1992 and an investigation by the NSF independent inspector general (which generally supported NSF decisions and actions) in 1993.

10. For this reason, there are a number of pilot projects being undertaken by cable and other providers apparently intended to gain a better understanding of consumer behavior and preferences.

11. One assessment, quoting several financial analysts, suggests that the combined winter rollback of cable rates by the Federal Communications Commission and the demise of the Bell Atlantic-Tele-Communications Inc. merger "will transform what had been a dynamic growth- and investment-oriented business into a more defensive cautious industry." It pointed to the prospect of difficulty generating the cash to invest in building the information superhighway and expressed skepticism about potential demand for consumer and entertainment-oriented services over it. Gilpin, Kenneth. 1994. "Market Place: A One-Two Combination Staggers the Cable Television Industry," *New York Times*, March 7.

On the telephone company side, Regional Bell Holding Company bonds lost some appeal as investments because of bond buyer concerns arising from the current turmoil and confusion in telecommunications (although some analysts suggest that those concerns may be overblown). Hardy, Eric S. 1994. "The Superhighway's Slow Lane," *Forbes*, March 14, p. 136.

12. The Internet does not include settlements and charges associated with the provision of universal access subsidies, which are found in telephony. On the other hand, consumer advocacy groups often allege that the magnitude and burden of some telephony cost elements are both hard to sort out and sometimes reflect arbitrary accounting conventions.

13. That larger community will embrace educators, researchers, and their associates through links to their homes as well as their employers or schools.

14. MacKie-Mason, Jeffrey K. and Hal R. Varian. 1993. "Pricing the Internet," prepared for a conference, Public Access to the Internet, John F. Kennedy School of Government, Harvard University, May 26-27, 1993, November 26 version. Note that there are also some labor costs associated with the support of those facilities and the relatively low cost of a Network Operations Center.

An illustration of how capabilities relate to costs may be found in experiences with NSFNET. The expansion from T1 to T3 service increased costs "dramatically," with $7 million of the $8 million in additional cost to upgrade service at the first eight sites attributed to the direct costs of transmission services (circuits, equipment, installation, and maintenance). "The three primary reasons for this cost increase were the new high-performance router technology, the costs of the additional capacity and speed on the MCI lines, and the requirement that the local lines [between internal and external switches] be able to handle T3 speeds." Office of the Inspector General, National Science Foundation, 1993, "Review of NSFNET."

15. Mile count of 16,200 from Jordan Becker, Advanced Network and Services, personal communication, April 3, 1994.

16. In the digital service hierarchy, the first step in this direction would be service at 1.5 Mbps.

17. Historically ISDN commanded a substantial premium, but that is changing in several areas.

18. There are also philosophical reasons in support of a private-sector approach, which was initiated under previous Republican administrations.

19. Snyder, Thomas D. 1989. *1989 Digest of Education Statistics*. National Center for Education Statistics, Washington, D.C., December; U.S. Department of Education, National Center for Education Statistics (NCES). 1993. *The Condition of Education, 1993*. NCES 93-290. NCES, Washington, D.C., June.

20. There is no general mechanism for guaranteeing that research (or education) applications that require the most advanced capabilities will be able to obtain them when or where needed. Regulations typically prevent the telephone companies from contributing (donating) the advanced new services while under tariff, and the window of opportunity for untariffed services is limited to very short periods for technology testing and experimentation. Those sites that do not have adequate connections in place will usually be unable to pay for them on an individual case basis as the charges will usually be much higher than when a planned buildout occurs for major portions of a community. Finally, sites that are in rural areas or those locations for which no good economic rationale can be made for a buildout may forever be disadvantaged without administrative remedies to offset the extra cost.

21. However, MIT also notes that its charges for connecting to the Internet via its regional network, NEARnet, could be an order-of-magnitude higher if either the nearest point of presence were in New York or the access fee were assigned to NEARnet members on the basis of usage. Other campus computing and communications managers queried informally by the committee presented similar degrees of uncertainty about costs. Personal communication, James Bruce, MIT, February 1994.

22. Obviously, the practicality of commercial Internet service that reaches most people's homes depends critically on the nature of the last-mile connection (see Chapter 2), in particular, the way in which the cable TV and local telephone networks evolve and interconnect. Unfortunately, recent plans and announcements have revolved around a form of interaction in which the traffic from customer premises requires only limited bandwidth. Thus there is a fear among Internet users that the new facilities will fall short of what is needed for Internet-like services and, by extension, for an optimal, effective NII.

23. Most cable TV networks broadcast analog video. They are not designed to make individual connections to homes and have no operating reverse channel. So interactive data services such as the Internet cannot use the cable TV network today in general.

24. See, for example, Continental Cablevision. 1994. "Continental Cablevision, PSI Launch Internet Service: First Commercial Internet Service Delivered via Cable Available Beginning Today in Cambridge, Massachusetts," news release, March 8. This service will involve a metropolitan area network on Continental's fiber trunk lines, which Continental characterizes as an "off-MAN subscription to the Internet." More generally, a variety of special modems is being developed for use in trials for data communication over cable; programs have been announced by Prodigy and Cox Cable Communications as well as CompuServe Information Service and America Online Inc. See "Online Services Announce Migration to Cable TV; Move Driven by Low-Cost Zenith Homeworks Modem," pp. 8-9 in *Information & Interactive Services Report*, December 17, 1993.

25. Since the committee generated this idea, it has learned that such a study may be requested through telecommunications legislation under consideration in 1994.

26. Remarks by Stephen Wolff during a National Net '93 EDUCOM conference, Extending the Benefits, held April 14-16, 1993, in Washington, D.C.

27. Whereas it can be argued that businesses and other for-profit user organizations may also seek institutional connections, they are more likely to be able to provide the local infrastructure necessary to distribute access within an institution to the individuals that need it.

28. McClure, Charles R., et al. 1994. *Connecting Rural Public Libraries to the Internet: Project GAIN—Global Access Information Network*, Project Evaluation Report prepared for NYSERNet, Inc., Information Management Consultant Services Inc., Manlius, New York, February 15.

29. For this reason, it is sometimes suggested that qualified outsiders be given access to institutions with adequate local infrastructure during periods of otherwise low use.

30. There is a trend toward more affordable Serial Line IP (SLIP) access for some local public institutions such as high schools.

31. McClure et al., 1994, *Connecting Rural Public Libraries to the Internet.*

32. The original authorization limit for NSFNET approved by the National Science Board was $14 million; it was increased in June 1989 to $20 million based on the increase in traffic from September 1987 (75 million packets per month) to February 1989 (600 million packets per month) and the increase in number of institutions connected to regional networks. In May 1990 the National Science Board approved an increase in the authorization limit to $28 million in connection with the upgrade to T3 service, a limit that is expected to be met. See Office of Inspector General, National Science Foundation, 1993, "Review of NSFNET."

33. The transition will be eased by the planned gradual decline in support for regional networks.

34. For example, MIT's internal costs may exceed its payments to NEARnet by an order of magnitude.

35. Fisher, Lawrence M. 1994. "Reining in the Rising Hidden Costs of PC Ownership," *New York Times*, March 27, p. F10.

36. The package included a Macintosh Color Classic computer, a LaserWriter printer, an external CD-ROM drive and cable, and a computer-fax modem for hardware; a multipurpose software package; communications software; and a graphics/video-handling package (QuickTime Starter Kit). McClure et al., 1994, *Connecting Rural Public Libraries to the Internet*, p. 7.

37. The prominence of local facilities in the Internet-based research and education context points to a fundamental contrast between the telephony and NREN models of communication: the telephone company model employs intelligence in the network (owned by the provider) and assumes "dumb" user facilities; the NREN model assumes a relatively "dumb" network and intelligent user facilities. These models are reflected in the approach taken to users or customers: somewhat oversimplified, the telephone company customer is a consumer, and the NREN customer is a participant (who, for example, puts his or her material on the network for others).

38. Pricing for information services is structured in different ways, and pricing structures are volatile. Some are more subscription oriented (one up-front fee covers usage over a period); some are largely by the transaction, although there may be additional start-up and/or annual or monthly fees. Transaction costs may be by the page (e.g., $1.00 to $2.00 for LEXIS-NEXIS) or by the record (e.g., $0.12 to $30.00, depending on the database, for Dialog). There are also charges for connect time in some instances, which may range from less than $1 to several dollars per minute.

39. According to Roger Noll, "Charging higher prices to users with more intense demands [for information products], and lower prices to users who might otherwise be excluded, maximizes the diffusion of the information product." Noll, 1993, "The Economics of Information: A User's Guide," p. 36.

40. Congestion control mechanisms inside the network can push back on the traffic source dynamically as congestion is detected inside the network. The Internet, which lacks any pricing controls, depends on these push-back techniques to control offered load and limit congestion. The techniques are imperfect, however. Not only is the congestion control algorithm not enforced by the network, but it is also not even part of the protocol architecture of applications that do not use TCP. UDP and video traffic have no current implementation of this, and in fact are "unfair" to the hosts that do implement congestion control.

41. PL 102-194, Section 102.

42. However, the analogy with a freeway system is imperfect, because it takes little

effort to set a computer on an infinite network capacity-consuming task, whereas highway use demands time from the driver.

43. MacKie-Mason and Varian, 1993, "Pricing the Internet."

44. MacKie-Mason and Varian, 1993, "Pricing the Internet," p. 16.

45. To some observers it seems that paying users will not be happy with best-effort service. Experience suggests that user satisfaction depends on the quality of service delivered, in which case it is argued that usage-based charging is the efficient approach. Unknown is how strong that relationship may be, and what it implies for consumer response to alternative pricing strategies.

46. See Shenker, Scott. 1993. "Service Models and Pricing Policies for an Integrated Services Internet," June 8 version, prepared for a conference, Public Access to the Internet, John F. Kennedy School of Government, Harvard University, May 26-27, 1993; and Cocchi, Ron, et al. 1993. "Pricing in Computer Networks: Motivation, Formulation, and Example," August 30.

47. In retrospect, pricing mistakes may have contributed to poor market performance for X.25 and ISDN offerings.

48. Interestingly, however, in the consumer market, Prodigy and GEnie were reported to have abandoned flat-fee for metered systems based on connect time in early 1994. Volatility in pricing appears to be the norm in network-based services, given imperfect understanding of nonconstant consumer behavior and its impacts on network load and on revenues. See, for example, Lewis, Peter H. 1994. "A Traffic Jam on the Data Highway," *New York Times*, February 2, pp. D1, D5.

49. The committee recognizes that price and cost often seem to have had a tenuous relationship in communications. For example, recent public discussion of the apparent costs of sending a fax by Internet versus over the telephone system points to the fact that the phone system allocates more costs more fully than does the Internet. Although the phone system entails some extra costs associated with mandatory cross-subsidies, the true magnitude of those costs is subject to debate, as is any particular allocation of costs.

50. CSTB, 1988, *Toward a National Research Network*.

51. Shenker, Scott, David D. Clark, and Lixia Zhang. n.d. "A Scheduling Service Model and a Scheduling Architecture for an Integrated Services Packet Network"; Cocchi, Ron, et al., 1993, "Pricing in Computer Networks: Motivation, Formulation, and Example"; and Estrin, Deborah, and Lixia Zhang. 1991. "Design Considerations for Usage Accounting and Feedback in Internetworks," IFIP International Conference on Integrated Network Management, April, pp. 719-733.

52. For example, those with fewer resources might have to rely more on longer-term prepaid arrangements, while those with more resources would have more flexibility to afford spontaneous and more expensive arrangements, and so on.

53. Specifically, there may be an analogy to the kinds of access charges seen in the telephone system that combine basic network access in the local loop with usage charges for long-haul or special services (e.g., the Exchange Network Facilities for Interstate Access tariff and its progeny).

54. Note that the High-Performance Computing Act of 1991 provided that "[a]ll Federal agencies and departments are authorized to allow recipients of Federal research grants to use grant moneys to pay for computer networking expenses." PL 102-194, section 102(f).

55. AT&T and other companies have been reported as favoring some form of targeted individual assistance delivered along the lines of "telephone stamps" intended to assure that benefits are provided to genuinely low-income people. Pearl, Daniel. 1994. "Debate Over Universal Access Rights Will Shape Rules Governing the Future of Communications," *Wall Street Journal*, January 14, p. A12.

56. MacKie-Mason and Varian, 1993, "Pricing the Internet," pp. 16-17.

57. The network is a general-purpose vehicle for many services. Information services are more specific, begging the question of which services might be eligible for coverage in a financial support program.

58. Einhorn, Michael A. 1993. "Toward Greater Achievements: New Policies for the Internet."

59. Noll, 1993, "The Economics of Information: A User's Guide."

6

Government Roles and Opportunities

The National Information Infrastructure (NII) initiative presents the federal government with an opportunity and an obligation to alter, enrich, and extend the existing elements of U.S. communications and information infrastructure. Pursuing that opportunity is an enormous undertaking, and the long-lived investments and interdependencies involved emphasize the need for effective up-front consideration of technical aspects and of striking a balance among many interests. Although the narrower National Research and Education Network (NREN) vision was captured in a specific program of public-sector activities, it was barely visible to many observers.[1] By contrast, the NII as envisioned by industry, public interest advocates, the research and education communities, the administration, and this committee is a far broader and more exciting concept that promises to alter and enhance activities across society and the economy. As yet, however, no coherent program or plan of execution exists for the NII.

It is necessary now to lay the proper groundwork for construction of a national resource whose complexity and significance for the nation for decades to come are only partly captured in the oft-used term "Information Superhighway." There is a clear opportunity for the federal government to act as a catalyst for wise development of the NII and as an arbiter among the various interests that must be balanced for the NII to serve a broad array of national needs. The federal government must effect a delicate balance between the free-for-all chaos likely to result from a hands-off posture and the overcontrolled bureaucratic process that can result from being too heavy-handed.

Moreover, the need for greatly expanded technological leadership—

of the kind that drove the NREN program—both in the development and use of network-related technologies has emerged with the NII initiative. This need is noted both in the 1994 report on the High Performance Computing and Communications (HPCC) program[2] and in the FY95 Clinton-Gore administration's budget proposals, whose introduction comments on the end of the Cold War and the growth in international competitiveness as motivations for a strong research effort.[3]

Both the NREN and NII efforts build on a history of Department of Defense (DOD) support for network-related technology and associated technology transfer; the Internet, in particular, constitutes a major technology transfer success story. The federal government now has a unique opportunity to build on that success through investment to further advance the underlying technologies (to support the technological underpinnings for the services that will ride over the network(s) and to connect users with the information they seek) and to develop quality information resources (e.g., databases consisting of government information or modules for educational curricula for which information infrastructure is a tool) that will further the use of the networks (see Chapter 4).[4] The Open Data Network (ODN) outlined in Chapter 2 provides a context for such investments; the emphasis on architecture, interfaces, and services provides a general and flexible framework that can accommodate many technologies, applications, and types of providers and users.

In broadening the policy focus from NREN to the NII, the federal government can play a variety of roles, each lending itself to a variety of mechanisms. Key roles, which are not mutually exclusive, include:

- Providing leadership and vision,
- Balancing interests and airing competing perspectives, and
- Influencing the shape of the information infrastructure.

These roles are considered in turn below. Given the expertise of the committee and the scope of this project, specific issues relating to regulation and deregulation are not examined in detail,[5] although the influence of regulation on competition and deployment of new technology is acknowledged.

LEADERSHIP AND VISION

Leadership and vision are hard to define and harder to deliver, but the history of both the NREN and NII initiatives underscores the value of both. The NREN program owes its existence to the vision of enlightened research funding-agency program officers and their congressional supporters. Given the flurry of private and public-sector activity in this area in the year and a half following the election of President William Clinton and Vice President Albert Gore, the NII initiative illustrates what can begin to happen when leadership at the highest levels is combined with vision. Fundamental to the administration's vision is the notion that an NREN program, per se, is not enough to meet either the needs of the research and education communities or of the nation at large; it must be combined with other elements from a broader domain to achieve those goals.

The administration has taken many steps to catalyze an NII. It has:

- Expressed its vision for the NII through speeches, white papers, and other policy statements containing high-level goals and principles (see Box 1.4 in Chapter 1);
- Created a new vehicle, a cross-cutting interagency Information Infrastructure Task Force (IITF) to explore areas where policy may need to be formulated or changed and to gather inputs from within and outside the government;
- Accelerated and expanded the use of information infrastructure within the federal government (for dissemination of government-generated information and services to the public, communication within the government, and communication between the government and the public); and
- Begun to use the federal budget process to define and redefine the national interest, notably by expanding the HPCC program in 1993 to include the Information Infrastructure Technology and Applications (IITA) component; launching the National Telecommunications and Information Administration (NTIA) Telecommunications and Information Infrastructure Assistance Program to advance and demonstrate options for connecting such public (and publicly financed) institutions as schools, hospitals, and libraries; and aggregating relevant programs relating to both research and development and deployment under a new NII category in the federal budget.

All of the above activities convey and reinforce a sense of importance and priority for information infrastructure. However, they address only some of the what and the why, and leave uncertain much of the how, when, and by whom, of the implementation of an NII. Aside from providing a statement of goals and vision, it is far from certain how the administration can lead or guide the private sector, on which the bulk of the NII investment hinges. The proposed legislative and regulatory changes[6] can set only gross parameters for private action; in fact, with respect to financing and control, the growing privatization of the Internet constitutes a stepping back from direct influence by government. Moreover, the broad reach of the NII, which includes organizations that generate, transport, and use information—and which may be necessary to ensure sufficient private investment—further complicates the problems of government coordination and effective leadership. It also raises questions about implications for the original constituencies of the NREN program.

If, as discussed in Chapters 2 and 4, the NII is to inherit the strengths of the Internet architecture and culture, an aggressive approach is necessary. However, such an approach leaves open the possibility of a research and education environment supported largely by the Internet, on the one hand, and a commercial, mass-market environment geared to entertainment services supported by a far less open network, on the other hand.

If the committee's vision of an open NII is to be achieved in a timely manner, it is essential that more than high-level goals and principles be articulated. Developing an ODN architecture requires significant departures from past patterns of private investment in infrastructure. Although technology can be used to enable an integrated infrastructure, economics tends to promote separation rather than integration, with investments and the development of capabilities driven by whoever is best capitalized. The availability of capital reflects the working of the marketplace, but the NREN program legacy addresses those groups for whom the market does not necessarily work well.

Fundamental to the challenge of providing leadership and vision is the committee's perception that there is a gap between the NREN and NII visions, with the NII even less defined than was the NREN. Specifically, the NREN program has focused on the physical wires, switches, and network problems; the NII focuses on a larger, more complex infrastructure involving people, processes, and information resources that exploit the network. *There is a unique role for the federal government in bridging that gap and in providing leadership.*

Leadership in Development and Deployment of Infrastructure

To bridge the NREN-NII gap, the federal government must address both the NII's architecture (its design and implementation) and deployment (which reflects pricing, marketing, support, subsidies, and resource distribution). As discussed below, this entails actions by multiple agencies. How the architecture is defined will influence deployment by determining the shape of products people can buy, their pricing and therefore affordability, and so on.

Key architectural issues include what features or characteristics should be a part of the NII, what services should be present, and what it means for a technology to be a part of the NII (see Chapter 2). Only part of the problem is the one of physical infrastructure, and here the analogy to a railroad system may be more apt than that of a highway, inasmuch as there is a problem assuring that the pieces will all fit together. Other parts relate to the capabilities the physical facilities will support, from communicating to doing things with information. Architectural and technology needs are not static; they will require ongoing research and development.

The goals for deployment are less precise than those relating to architecture. Key issues relating to deployment include financing, control (of facilities and services), and their interrelationship; these factors may all vary over time and affect the timing for the achievement of deployment goals. An essential role for the federal government in deployment is to assure access to infrastructure, adequate maintenance, essential technical services, a clear migration path, and technological development.

The NSFNET-Internet transition concerns the research, education, and library communities specifically with regard to its financial implications. Here there must be a recognition that the information infrastructure will fail to reach its potential if in its development we forget these key communities. *There is a need for guidelines and oversight to make sure that we take care of NREN constituencies—the research, education, and library communities—during the transition.* Some targeted financial assistance may be needed, as recommended by the committee in Chapter 5. Further, the full economic benefit of the NII will occur only when all organizations, including small business and government at every level, become part of the infrastructure, as well as most individuals; an NII can constitute an information bazaar, with powerful positive economic and social consequences.[7] However, achieving this scope may take a concerted effort at local and state levels, as well as at the national level.

Development of infrastructure at the state level, driven in part by an interest in achieving greater efficiencies for state government functions (government as a distributed enterprise) and in part to support local economic development, will contribute to the NII as it relates to the K-12 education and public library communities (in some cases, for example in North Carolina, state networks serve the research community in addition to delivering various government services); somewhat less certain in terms of scope, scale, and staying power are regional networks. State and regional efforts may affect state regulatory actions bearing on technologies in the "last mile" (some of which may be influenced by potential legislation and efforts by the Federal Communications Commission (FCC)), the emergence of greater competition in the local exchange, and the possibility that more than one physical plant (e.g., cable and telephony) may be upgraded).[8] See Appendix D for an overview of state and regional developments and issues.

Illustrated below are opportunities for the federal government to influence the broader complex of federal, state, and local actions in the area of education. These opportunities apply equally well to research and libraries.

Leadership in Education

Within the context of nurturing an NII, one area appears to call for a focused leadership effort as well as additional resources: education. The administration's goal of connecting all classrooms by 2000 and various industry programs' support for the use of information technology in schools are helpful, but a consistent, focused effort is warranted in this area at a time when many perceive much promise but too few financial, technical, and human resources to realize it. A clearer and more effective effort within the Department of Education would constitute an important first step.

The Department of Education is in principle the best federal entity to bring together considerations of access, content, and support for K-12 education. Although actions taken by others, including industry, may enhance access and connectivity, the timing and consistency of their efforts are uncertain. *Moreover, access and connectivity are necessary but far from sufficient: the integration of networking into K-12 education requires that mainstream educational services be available over the network.* Providing such services, in turn, requires a highly organized effort to create the software, secure state government and community approval of materials and instructional environment, train thousands of teachers, and provide hands-on assistance to teachers confronting new applications.[9] Integrating networking into education also requires an ongoing program of research into the design and implementation of infrastructure technology and applications for education, because the problem of providing and supporting access to network-based resources in education is larger and also contains more unknowns than support for research networking. The introduction of the IITA component into the HPCC program has infused resources and talent into this research area, although the proliferation of educational applications activities among research agencies raises questions about direction and the potential for duplication of effort.

Achieving a specific locus of responsibility and accountability within the Department of Education (ED) will be an essential first step. Consistent with this objective, the administration has appointed a respected individual as special advisor on technology in the office of the deputy secretary of education. The committee recognizes that progress and effective action at ED will require a much greater level of technical expertise and experience than has been available at that agency. Proposed legislation (S-1040, the Technology for Education Act of 1994) would establish an Office of Educational Technology headed by a director of educational technology reporting directly to the secretary of education. That bill would provide federal funding for education planning, equipment purchases by disadvantaged schools, educational technology research, and development of educational software, as well as grants for local school districts planning to incorporate technology into education. Until sufficient internal technical competence is established, ED should build on relevant programs at the National Science Foundation (NSF), the Department of Energy (DOE), the National Aeronautics and Space Administration (NASA), and other agencies involved in the NREN and HPCC programs that have explored network applications for science and math education.[10] In addition, closer coordination with the Department of Commerce (DOC), which has been assigned a leadership role in the administration's NII activities, could help to better and more systematically engage industry in this area, building on the broadening base of voluntary action emerging from industry.

RECOMMENDATION: K-12 Education

The committee concludes that there is a clear and present opportunity to improve K-12 education by the integration of networking into the U.S. educational system. Consistent with recent legislative proposals and the selection of education as one of the emphases in the National Information Infrastructure initiative, the committee recommends the following:

—*The federal government, through the Department of Education, should take a leadership role in articulating to other federal agencies, state departments of education, and other members of the education community the objectives and the benefits of networking in K-12 education. It should define a national agenda that can guide efforts at the state and local level.*

—*Since this leadership requires technical competence, the Department of Education should, in the short term, pursue collaborations with the National Science Foundation and other research agencies, but in the long term should acquire internal technical expertise at a sufficiently senior level.*

—*The Department of Education should set an aggressive agenda for research on telecomputing technology in education. This research should address benefits and applications of high-bandwidth communication and services and the transfer of related technologies to educational applications.*

—*The federal government should continue, and if possible expand, federal funding through matching grants, leveraging state, local, and industrial funds, to stimulate grass-roots deployment of networks in the schools.*

BALANCING OF INTERESTS

Effective actions undertaken by the federal government must build on a consensus on goals, values, and the balance of public and private interests. Toward that end, the federal government can play an important role in gathering input from all segments of society and the economy and in weighing and balancing different interests. An objective of the balancing process will be the critical assessment of the alternative NII visions offered or implied by different groups and options for achieving the effective integration of those visions, as outlined in Chapters 1 and 2. The nature of the process—how adequately different perspectives are treated and how the process is structured—will determine its effectiveness.

The move from an NREN to an NII focus broadens the set of involved agencies and constituencies. In the short term, there is a sense of chaos resulting from the proliferation of many parties with competing agendas, but the process is expected to resolve and reconcile many differences over the long term. In the meantime, the expectations of the electorate are being raised. A benefit from that broadening of input is a fuller national consideration of competing interests and needs relating to information infrastructure. This fuller consideration is especially important to support decisions relating to societal equity—including access to networks and the information resources available on them (see Chapter 4)—and federal budget allocations (see below).

Diverse and Fragmented Public and Private Interests

The growth of networking in public and private contexts and the development of an integrated NII involve a number of entities whose roles are changing. They include the federal government; the state governments; schools, libraries, universities, and other educational institutions; regional and other mid-level network providers; and the commercial sector, including both providers and users of network-based services and other nongovernmental organizations.[11] There are thus a large number of stakeholders that include, or will be affected by the actions of, the major entities that will most directly shape elements of the information infrastructure.

The proliferation of stakeholders, the rapid growth of the Internet, and the prospect of broad interconnection among different kinds of network infrastructures raise many questions about the Internet, in particular. There is a shift from a voluntary community that has effectively run the Internet to a set of more formal and informal organizations (including government agencies, telecommunications companies, and public interest groups) that want to participate.[12] Some want only to have access to the emerging NII at an affordable cost. Others want a say in the development and operation of the Internet and/or the larger information infrastructure. And some would like to control the entire enterprise, including the transport network plus the information services available on the network.

Several entities are advancing plans relating to information infrastructure. Prominent among them are the administration's cross-agency IITF; the multiagency High Performance Computing, Communications, and Information Technology (HPCCIT) subcommittee (under the Office of Science and Technology Policy (OSTP)-National Science and Technology Council (NSTC)-Committee on Information and Communication (CIC) R&D umbrella), for which a long-awaited advisory committee may eventually provide industry and academic input; the multiagency Federal Networking Council (FNC) and its associated advisory committee (FNCAC); federal mission agencies whose programmatic needs drive network implementation in their portions of the NREN program; and an assortment of private entities, including trade, professional, and advocacy groups, such as the Council on Competitiveness, the Telecommunications Policy Roundtable, the Coalition for Networked Information, EDUCOM, the Electronic Frontier Foundation, Computer Professionals for Social Responsibility, the Computer Systems Policy Project, the Internet Society, the Cross-Industry Working Team, and so on, as well as direct representation from the entertainment, cable, telephone and other telecommunications, and information-providing and publishing industries.[13] This existing set of involved parties is bewilderingly large and diverse, and it is growing. It is also fragmented, with most entities focusing on specific sets of issues or perspectives.

A consequence of the broadening and fragmentation is that the concerns of the research, education, and library communities are not consistently addressed and are in danger of not being heard. Events to date suggest that these communities are barely present at the table for key discussions, notwithstanding the rhetoric about serving public interests and the political appeal of investing in education. The risk that the research and education communities may be isolated or underrepresented is magnified by the prominence of players in the infrastructure arena that have minimal if any historic relationship with these communities (notably the entertainment, cable TV, and commercial software firms that have been prominent in the recent flurry of mergers, alliances, investment programs, and new service announcements). With so many stakeholders and with such an emphasis on achieving a broad base of private investment, it is not surprising that the recently established NII Advisory Council has negligible representation from research and education (although perhaps it is inevitable that such a high-level advisory committee would have only token representation from the broad set of constituencies).

However, it is critical that the federal government not back away from the research and education communities. It has achieved successes in research that should be reinforced and built upon, and based on that experience the committee offers the value judgment that the potential benefits for education at all levels justify a focused, sustained effort to develop and use information infrastructure in that arena.

To assure that federal budget allocations reflect both efficiency and equity considerations, it is desirable that there be a broad balancing in the choices made between research and education and other kinds of programs; within research and education for information infrastructure and other kinds of input; and within research for network-related versus other kinds of research and development. Early debates over the NII make clear that fuller consideration makes policy analysis more complicated, increases the risks of politicization, and aggravates the problems of coordination within the federal government (and among federal, state, and local levels).

Coordination and Management

The scope of the NII initiative introduces considerable problems of coordination, control, and accountability within government even if the government's role remains contained. The evolution of the narrower NREN program illustrates this problem. Although there is an interagency communication mechanism in the form of the FNC, and its associated FNCAC, as well as the larger HPCC coordination processes,[14] this committee is concerned that there is no truly effective mechanism for coordination of NREN efforts among agencies that is guided by input from the research and education communities. Some of the uncertainty about the collective future of NSFNET, ESnet, and the NASA Science Internet reflects their different degrees of visibility across the research and education communities: NSF (and its networking efforts) is far more visible across the board. Meanwhile, the introduction of new federal programs serving NREN communities (such as NTIA's Telecommunications and Information Infrastructure Assistance Program) increases the difficulty of coordination within the federal government alone, not to mention between government, industry, and academia. The shift to an NII emphasis only exacerbates this problem.

Although specific current federal efforts to promote an NII are hard to pin down, the centerpiece is clearly the IITF. The IITF appears designed to provide a policy framework for meeting general public infrastructure needs, attacking the broader nature of the NII challenge, interacting with many stakeholders, and serving as a vehicle for coordination and communication across the government on several interconnected policy issues. The current focus appears to include lowest-common-denominator approaches to the most basic connectivity or industry-directed provision of services to households. However, as now empowered, the IITF neither fully embraces the research perspectives that have grown with the HPCC and NREN programs nor satisfactorily addresses the education and library communities' needs for more than minimal access. By contrast, the vision of the Open Data Network articulated by this committee in Chapter 2 is not a lowest-common-denominator approach: it incorporates a need for an evolving low end, but it lays the foundation for a richer construct for the future. *The challenge for the country is to shape the architecture of the network so that the NII that results meets not just short-term commercial objectives, but also longer-term societal needs.*[15] *It is important to appreciate these differences in outlook now, since progress dictates that rough agreement on an NII vision be achieved sooner rather than later.*

As this report is being written, the IITF is divided up into committees and working groups, with an overall focus on telecommunications and information policy to enable the NII to meet broad social and economic objectives. The IITF serves an important function, but from the perspective of realizing the ODN envisioned, it raises three concerns: (1) by design, it is nontechnical; (2) it has the strengths and weaknesses of a cross-agency entity; and (3) it is an evolving construct with an uncertain future.

Uncertain Technical Expertise

The technology component of the IITF appears confined to a relatively new Technology Policy Working Group (TPWG), which is charged with addressing issues of technology policy. Officials distinguish this mission from the research and development coordination activities and responsibilities of the HPCCIT, which has overseen the NREN program and other elements of the HPCC program.[16] The TPWG is at least potentially a bridge to the technical expertise and agency representation within the HPCCIT and other components of the National Science and Technology Council (specifically, the CIC) under the OSTP. For example, it appears to involve some of the same individuals, at middle- and upper-management levels, as the HPCCIT. This linkage is valuable, because the people with experience in new network-based technologies and applications are those who have been using it—notably, individuals at research-funding and certain other mission agencies. However, the TPWG is a small component of the IITF, and its ability (or, indeed, the ability of the NSTC) to bring essential technical expertise to bear on NII policy formulation appears uncertain. This uncertain connection is a grave concern, given the need for ongoing input into evolving architectural, deployment, and associated technology development dimensions of the NII. Since developments over the next several (formative) years will require more creativity than consolidation, realizing the NII potential envisioned by the administration and the committee will require continuing involvement of players with the best technical understanding of what the issues are. Thus the committee is concerned lest the policy pendulum swing too far from the technology-oriented programs and competences that have been a major strength of the NREN program.

In addition to the NII's intrinsic dependence on technology (see "Influencing the Shape of the Information Infrastructure," below), the IITF's current regulatory and legal emphases themselves have technical dimensions. For example, the policy objective of universal access can be met only with a characterization in technical terms of what that access is. Hasty or fragmentary actions without the benefit of informed technical insights risk being ineffective or counterproductive. The quality of the input and deliberations relating to both architecture and deployment will be critical.

Cross-agency and Uncertain Structure

The need to blend multiple perspectives has given rise to a variety of interagency structures.[17] However, it is hard to lead by committee; existing and recent cross-agency bodies tend to have provided communication and coordination functions (important as they are) at best. *A structure with more permanence, responsibility, and accountability than an advisory committee, coordinating council, or task force—combined with ongoing support from the offices of the president and vice president—is needed to sustain a dynamic NII development process.*

Moreover, the committee is further concerned that once the IITF has accomplished its mission, any ongoing oversight and coordination role may devolve to a single existing agency (such as the DOC or one of its components, the National Institute of Standards and Technology (NIST) or the NTIA,[18] or, as suggested by the draft report of the Federal Internetworking Requirements Panel (FIRP), the Office of Management and Budget (OMB)).[19] The concept of a greater OMB role is intriguing because of the cross-cutting nature of OMB's scope, but the agency's mission and emphasis on fiscal conservatism, while important for the tough trade-offs that an NII initiative can engender, seem at odds with the need to promulgate a visionary program; integrate technical, economic, social, and legal perspectives; secure broad involvement across government, industry, and academia; and attend to other technical aspects of the NII challenge. Also, the agency has not been a major user of networks itself, implying limited experience and insight in this arena.

Yet the designation of OMB is symptomatic of the problem of finding an appropriate, effective institutional focus: because mission agencies dependent on networks have been frustrated with traditional procurement processes and institutions, the FIRP apparently sought an entity with some access to technical skill, some relation to actual network users, and the capability or authority to identify and assign actions or roles in support of government networking. The FIRP's draft recommendations attest to the problems of aligning experience with charter. Given the inherent limitations of the mission orientation, key constituencies, experience base, resource constraints, and so on characterizing individual agencies, *the committee believes that no one existing agency can play the broad and ongoing role envisioned.*

Nevertheless, specifically expanding and empowering the Advanced Research Projects Agency (ARPA) and NSF to contribute more broadly to the NII could assure a dynamic outlook on the underlying technology and architecture as well as prepare for the ongoing process (including specification, implementation, and support) of upgrading that technological base. NII creation still presents leading-edge research challenges and is very interdisciplinary, arenas in which ARPA, in particular, has succeeded, and to which NSF can contribute substantially through its support of both network research and network-based research.[20]

RECOMMENDATION: Leadership and Guidance

The vision of a national information infrastructure (NII) as articulated by the administration emphasizes significant U.S. social and economic concerns but leaves largely unaddressed a number of critical technical issues. The technical roots associated with the NREN program and other components of the larger HPCC initiative must be effectively and consistently factored into that vision.

The committee recommends that the federal government expand its NII agenda to embrace the Open Data Network (ODN) architecture as a technical framework for the design and deployment of the NII. Required is a stable mechanism to provide the following:

- *Continued federal leadership in stimulating the development and deployment of an ODN architecture for the NII, integrating the technical, economic, and social considerations basic to achieving a truly national U.S. networking capability.*
- *Continued federal involvement in the development of standards for the NII. The committee does not conclude that the government should set the standards, but rather that it should support and participate in the ongoing standards-setting processes more effectively, bringing to those*

processes an advocacy for the public interest and for realization of an open and evolvable NII.

To this end, the committee further recommends that the federal government designate a body responsible for overseeing the technical and policy aspects of the evolution of the NII and its applications.

The Information Infrastructure Task Force (IITF), which focuses on policy issues, is not sufficient for this role; from the perspective of realizing the ODN architecture, it raises three concerns: (1) by design, the IITF focuses on nontechnical issues and is dominated by nontechnical perspectives; (2) it has the strengths and weaknesses of a cross-agency entity; and (3) it is an evolving construct with an uncertain future. The National Science and Technology Council (NSTC), and its component Committee on Information and Communication R&D, which oversees the High Performance Computing and Communications Information Technology activity, is also not sufficient for this role; it raises these concerns: (1) its mission is to coordinate R&D programs, and (2) it, too, has the strengths and weaknesses of a cross-agency entity.

What appears to be needed is a body that will effectively blend the technical competence of the NSTC with the policy capabilities of the IITF and be able to function for the extended period of time required to develop and deploy an NII with an ODN architecture.

INFLUENCING THE SHAPE OF THE INFORMATION INFRASTRUCTURE

The federal government can influence the shape of the NII in terms of both architecture and deployment. In both instances, standards, procurement, regulation, and investment incentives are key mechanisms. This report focuses on standards and procurement, in addition to related research investments, as tools for shaping NII architecture and deployment; full consideration of regulations, which are effectively a more formal approach to standards, and investment incentives was beyond the scope of the committee.

Although most of the public debate over the NII has addressed issues specific to the U.S. context, the NREN program has demonstrated the benefits of easy international connectivity, including international information sharing, collaboration in research, and educational exchanges at all levels; it has also illustrated how difficult it can be, in some parts of the world, to achieve even physical connectivity. Expanded international interconnection will require bilateral and multilateral agreements, involving the Department of State, other agencies, and perhaps other bodies, most likely building on existing and prior law (although physical implementation will be effected with private investment[21]). See Appendix E.

Even more importantly, the prospect of broader international connection underscores the need to address issues that will arise with information-oriented applications, which will be affected by differences in legal regimes, values, and so on. Intellectual property rights, transborder data flow,[22] privacy, and security are among the areas that will present challenges for the international information infrastructure, challenges that U.S. information policy making should anticipate.

Influence on Architecture and Standards

As discussed in Chapter 2, the committee's vision of an Open Data Network entails achieving a more general and flexible architecture than appears likely to emerge independently from private-sector actions. In part for this reason, a central activity will be the development of appropriate standards and guidelines. *The process of setting standards is the only way that a high-level vision of the NII can be translated into a useful deployed infrastructure.* This report attempts to sketch a vision, but clearly this vision is partial and must be translated into a concrete architecture and a set of defining standards. Thus defining the vision and creating standards must be to some extent interdependent. Of course, setting standards is not the same as getting them adopted. However, the history of the Internet standards-setting process, characterized by vision and leadership by ARPA program managers and creativity among those in the research community that they funded, illustrates that sometimes the two activities can go hand-in-hand.

The committee sees the involvement of the government as critical in shaping future network standards. If the NII is to succeed, the government must stay involved in the process and find some better way than now exists to represent the broad interests of society in the standards-setting process.

Setting standards for infrastructure involves a broad range of entities with different competencies, constituencies, time scales, and effectiveness, all of which interact in a context in which standards setting is largely voluntary. Thus, part of the jurisdiction lies with the Federal Communications Commission, part with domestic voluntary standards committees (the American National Standards Institute, Institute for Electrical and Electronics Engineers, and so on), part in international bodies (the International Telecommunications Union (ITU) Telecommunications Sector (formerly CCITT), part with such voluntary groups as the Internet Engineering Task Force (IETF), and part with a variety of ad hoc and more formal industry consortia. The situation is complicated by the fact that, particularly in areas such as information infrastructure, U.S. actions must relate to a larger, international standards-setting process.

No organization at the moment holds the charter to set a global vision of the NII. The Internet Society represents one effort to provide coherence in this dimension. It reflects a bottom-up grass-roots approach that has so far marked the growth and evolution of the Internet; it is also moving toward more formal, liaison relationships with the International Organization for Standardization and the ITU, steps that would enhance its involvement in international standards setting (although within the Internet Society, the IETF has traditionally focused on lower-level protocol and architecture issues, and as characterized in Chapter 2, upcoming challenges relate to the middle and higher levels). Since the current broad base of stakeholders precludes direct control by the government, the government must decide what organizations it will support to bring into existence a vision for the NII as well as the supporting standards, and it must work internationally to establish the working relationships and the mandates that can make the NII a reality.

The committee is not recommending that the government charter one of its standards-setting agencies, such as NIST, to directly set or mandate all of the standards anticipated for the Open Data Network. Indeed, past attempts to influence the process directly in this way have not been effective. The attempt to force the use of Open Systems Interconnection (OSI) protocols by the promulgation of a federal government version, GOSIP, must be seen as a misguided attempt to exercise a governmental mandate. In the commercial marketplace, the contest between the OSI and TCP/IP protocol suites is over: the OSI market has largely disappeared,[23] and vendors who invested enormous sums in trying to develop this market are understandably upset. The difficulties of both abandoning previously chosen directions and deciding on standards for future directions are illustrated by the winter 1994 controversy over the draft report of the FIRP, which suggested that NIST abandon its position mandating the procurement of technologies implementing the OSI suite.[24]

Against this backdrop comes an administration effort to strengthen the involvement of NIST in the NII initiative. This can be seen in the significant expansion proposed for the FY95 NIST budget, the prominent role of NIST's director in the IITF activities, and the FIRP's draft recommendation that NIST identify federal preferred standards profiles and aim "to converge the Government to a single interconnected, interoperable standards based internetworking environment."[25] NIST manages the development of Federal Information Processing Standards (FIPS), and the FIPS system would be a vehicle to promote NII-compliant technology as suggested in Chapter 2.

The issue of a more active government role in setting standards is controversial. Many standards are being set in industry, particularly at the applications level (such as data standards emerging from the PC applications software environment). Many in industry believe that the ad hoc bottom-up standards-setting process that has characterized the U.S. computer industry and Internet context has been key to today's global leadership in those arenas. There is also concern in the business community about the ability of government officials to make the right choices (whether for standards or regulations), a view captured in a *Wall Street Journal* editorial contending that "it is truly hubris for these politicians to think they can somehow fine tune or stage manage the rapidly developing world of advanced technologies."[26] Even within government, opinions differ as to the appropriate timing and direction of standards setting. For example, a Federal Communications Commission official participating in a forum on wireless communications observed that the FCC preferred encouraging to mandating standards in a new industry, while an NTIA official was quoted as asking whether the FCC should do more than provide encouragement.[27]

On the other hand, standardization has been immature and conflict or lack of consensus has been apparent in such cross-cutting concerns as management, security, and network naming, areas where industry-driven standardization may be neither sufficient nor sufficiently timely. Moreover, the objective of providing a truly national, consumer-oriented set of services increases the decision-making stakes because consumer-oriented standards tend to be slow to change—the consequences of these decisions are evident for relatively long periods of time.[28] These are among the factors arguing for explicit attention to the direction, degree, and consistency of standards-setting actions across government.

Influence Through Procurement

Among the vehicles for promoting adoption of the ODN architecture is government procurement of relevant, NII-compliant technology for its own uses. Two kinds of procurement are at issue: procurement for the conduct of government as an enterprise, for which FIPS (and in mission-related defense contexts, military standards) are promulgated, and procurement of "research networks" that combine service to both internal and external parties with exploration of advanced technologies. This latter approach has been used successfully in the NREN program, which has both demonstrated technology and stimulated associated market development; it is consistent with the goals of the National Performance Review for broader use of network-based technology in the conduct of government activities; and it is compatible with the preliminary recommendations of the FIRP.[29]

The impending transformation of the NSFNET illustrates both how the relative federal government role has changed in the supply of infrastructure and how the need continues and expands for the government to support the research and education communities through effective communication about plans and prospects as well as appropriate delivery of financial assistance. Two key sets of issues relating to procurement are discussed below: influence on the Internet, and the fit between government approach and kind of network.

A third set of issues is subordinate to both of these other sets; it relates to the specifics of approaches chosen by individual agencies to meet mission needs. Today, NSF, DOE, and NASA each operate or contract for the operation of a dedicated backbone network providing services such as file transfer, electronic mail, and remote resource access. When the federal backbone efforts began there were no commercial providers capable of offering required services. Now components of the Internet, agency backbone networks have served as testbeds for experimentation with network services, algorithms, software, and hardware.

These agencies have taken different approaches. For example, NSF has developed cooperative agreements that delegate operation of a dedicated network (NSFNET and ultimately the vBNS) that satisfies a specified performance goal. DOE has attempted to procure, for an enhanced ESnet, telephone company facilities involving specified technologies (including asynchronous transfer mode (ATM)), in anticipation of commercial offerings; it has proposed taking a virtual private network approach. These arrangements have been made as separate logistical activities, based on the assumption of more or less separate user communities.[30]

The variation in approach across agencies raises questions about coordination. However, subject to the concerns discussed below, it is not obvious that any one approach is inherently superior unless it can be established by engineering-economic analysis that there are clear economies to merging these efforts.[31]

Influence on Future Oversight of the Internet

The federal government (and specifically the NSF) role has diminished to the point that it no longer appears that the government can simply "turn off" the Internet by removing payment; the government-funded portions have shrunk compared to other portions as the entirety has grown. With the trend to commercial networking and more indirect governmental involvement, the use of infrastructure procurement decisions to control and plan the Internet's growth will end. This is a critical problem in the transition to commercial services. Without some other means to provide overall guidance for Internet planning, chaotic growth may effectively disable the Internet and prevent future success.

The federal government is still seen as providing overall guidance and thus has the power to pass this role along to a successor in a coherent manner, or to retain it. However, the government's position is rapidly eroding with the increase in commercial and foreign country interests in the growth and use of the Internet. As a result, the government has a greater chance of influencing the succession of power than of wielding overall control over the Internet. The advent of the Internet Society—in the context of declining and uncertain federal support for the Internet plus the broadening of international interest in the Internet—points to the fact that there are alternatives to direct government involvement; those alternatives and their ramifications should be fully considered rather than be allowed to emerge by default.

The committee thus sees it as an obligation of the federal government to leave a clear line of succession for the oversight of the Internet. This is an essential element of planning for the larger NII. The committee concludes that a fight over ownership of the Internet architecture, which could easily occur in the power vacuum left if the federal government were to withdraw further, would be intolerably destructive, jeopardizing the future role of the Internet as part of the foundation for the NII.[32]

In today's unregulated and competitive world, monopoly is not a preferred tool to achieve coherence; the committee is suggesting neither that nor completely centralized control. But it is both appropriate and necessary for the government to stabilize those aspects of the overall Internet environment that permit the Internet to function. At the present time, many commercial providers have been effective at meeting among themselves to shape their part of the Internet, in part through the Commercial Internet Exchange (CIX). The CIX is not, however, suitable to take on the role of overall planning for Internet growth. It is properly seen as representing commercial interests, which are not the only voices to be heard. Also, the CIX involves only some commercial providers. Thus, to make the CIX the overseer of the Internet would seem to some to put the fox in charge of the hen house. The same argument about bias could be raised against any business consortium, trade association, or equivalent.

The second NSFNET solicitation and award (revolving around the vBNS; see Box 1.3 in Chapter 1) recognizes the need for continued centralized planning and contains two key components that address this need. One is the designation of an organization to manage Internet routing (the routing arbiter). The other is the topology implicit in the vBNS proposal itself, which provides network access points (NAPs) as a means to interconnect future service providers.[33] Whether these aspects of the solicitation are technically correct (and there is some criticism of them from commercial providers), it must be understood that this is the last time that the government can take such a step. *The committee does not believe that in the future a government solicitation will be an effective engineering means to shape the Internet. With this second NSFNET solicitation, the government is effectively out of the business of overseeing the Internet.*

Influence on Network Deployment and Technology Development

It is important to recognize the benefits of direct governmental involvement in network deployment. First, it accelerates the advent of wide-scale deployment to the point that the benefits of network attachment are real and visible to the users. Second, it drives the development of technology, which might not otherwise come into existence, since there is no initial market for it.

This first goal, accelerated deployment, has succeeded to the point that, for operational networks serving broad sectors of the community, the role of the government is already changing to that of a facilitator. This is an appropriate trend and should continue. By the indirect use of grants or subsidies (e.g., to institutions in support of network utilization), the government can accomplish the policy objective of rapid network deployment and at the same time encourage the development of private-sector network offerings. For example, because the Internet has evolved its own technologies for switching and routing, it has achieved faster deployment of new technology than is available publicly in the public switched network overall.

The second goal, technology development, will still justify direct investment in specific cases. For example, there is a government role in funding basic research related to architecture, because architectures having the shared nature and the scale of the NII class are not likely to be funded by industry; the same holds for research relating to the generality of solutions and level of integration of resulting architectures. The government is funding research in very high speed experimental networks, including joint efforts among government, industry, and academia to promote technology development in areas such as ATM and SONET (both of which reflect substantial industry R&D).[34] This is effective and should continue. The other option for technology development is at the other end of the performance scale, that is, the development of very inexpensive interconnection technology for schools and similar facilities for which the objective is wide penetration at low cost.

A significant component of public debate, including debate in connection with legislative language, over how and how much the federal government should support networking infrastructure has revolved around a model that divides networks into two categories: *experimental* networks, which are vehicles for testing out substantially new technologies and/or applications, and *production* networks, which use mature technologies to deliver services. Although this distinction appears reasonable on the surface, in practice it may be flawed.

Support for Experimental Networks

At any point in time, there will be a range of network technologies, from the very experimental to the commercially provided production networks. Today, all-optical networks are very experimental, ATM networks are emerging from the experimental realm, and Internet-style packet switching is supported by very mature products. At another time, what is considered experimental, mature, or in the middle will differ, but the range of options will always exist. The two ends of the spectrum are clear, but the middle is more complex.

Highly experimental networks, which are not yet of any operational benefit, are most properly funded by direct research grants from government and industry. They are characterized by high risk, they implement precompetitive technologies, they generate insights that can be widely shared, and, as a result, there is limited commercial incentive for any one company to undertake such projects. The current gigabit testbeds are supported by government grants from NSF and ARPA and by direct industrial support, especially for the transmission facilities. Other testbed programs appear desirable (see Chapter 2). For example, it is possible to imagine a testbed supporting access to schools that drives the development of very inexpensive interface equipment for voice-grade packet switching. Another possible experiment would involve wireless interconnection of rural locations.

Approach to Operational Networks and Intermediate Technologies

At the other end of the spectrum, for networks that are operational and that do not represent an advance in the current state of the art, it seems reasonable to have the users pay directly, as they might pay for telephone or cable TV access. Direct government funding of such networks seems inappropriate, although government subsidies to research and education users may be desirable (see Chapter 5).

There are, however, some gray areas that require consideration. One issue, related to operational networks, has to do with who does network integration. A network can be built by purchasing low-level components, like trunks and routers, and hiring staff to build these into a service. The alternative is to purchase the desired service directly. In the latter case, one could purchase the service from a general service provider, as in the purchase of telephone service, or one could contract to have a special service implemented, as in the case of the federal telecommunications system, FTS-2000.[35]

Another gray area concerns network technology that is somewhere in the middle of the spectrum between the purely experimental and the purely operational. At any time, there will exist networking concepts that are by no means yet a commercial commodity but that have matured to the point that it is reasonable to let a community of advanced application users exploit them. *It is very important to encourage the use of these intermediate technologies, first to help the selected user community, but more importantly to help prove the viability of the concept and push it toward commercialization.* The lessons of the past decade affirm the value of such proofs of concept. In this case, even though the network technology is not strictly experimental, the government may play a key role in purchasing and providing such a service to a selected user community (presumably a community for which there is a preexisting rationale for government support, such as research or education or a community involved in the pursuit of a specific and well-defined agency mission). For example, DOD was the principal customer for early implementations of ATM and SONET network technologies; it was able to afford the high cost, and its investments helped to lower the costs, facilitating the commercialization of these technologies.

There will always be a tension in this middle region, because as technology becomes more mature the commercial sector will show increasing interest in offering the service. Direct government provisioning may be seen as inhibiting commercial development; this is the allegation raised in criticism of the NSFNET arrangements during the early 1990s, when commercial Internet access providers emerged and alleged that government subsidies undercut them. These arguments continued into 1994 in the context of proposed legislation aimed at nurturing information infrastructure.[36]

The arguments raised by the commercial providers seem to suggest that this middle region of maturing technology does not exist. Government has been urged by carriers and other commercial service providers to divide all networks into research networks and operational networks, with a sharp demarcation between them.[37] This crisp distinction is not realistic, and it potentially hurts the development of evolving technology. Important new applications may not come into existence unless the infrastructure is there. Moreover, government-initiated applications tend to provide the creative requirements that bring to the surface hidden or latent demand. Even if the technology demands of maturing forms of infrastructure do not themselves mandate direct government support, demand for the timely development of applications may require it. *Therefore, the committee urges that the government not be restricted from purchasing intermediate and advanced network technology and providing it to selected user communities, so long as policies are in place to direct the technology's eventual migration to fully commercial services, and to ensure a degree of competition in the provisioning.*

Research and Development

There will be expanded opportunities (and, ideally, resources) for federal investment to further advance the technologies underlying information infrastructure. A strong government role will be valuable in the uncertain transitional period when it is unclear what NII markets will support and how commercial firms will act.

The NII will become a reality *if* the research conducted over the past three decades by industry and universities is continued. This research, much of which has been supported by the federal government, has addressed a wide variety of issues of substantial importance, including technology development and prototyping, fundamental science, and long-range, high-risk studies.

Networking research was first funded by ARPA beginning in the 1960s, expanding to NSF and then NASA, DOE, and other parts of the DOD in the 1980s (see Appendix A). This research has laid the foundation for the current Internet and helped to establish a multibillion-dollar industry based on this technology. The U.S. leadership in data network technology seems directly related to its preeminence in research in this area, and the committee would urge a continued program in experimental network research for this reason. It is easy to conclude, from the great success of the Internet, that all the necessary research has been done. Such a conclusion would be incorrect and very destructive.

Today, the multiagency HPCC program supports a wide array of research projects that will help the country move by the next century to a new, ubiquitous, service-rich communications and information environment. An important characteristic of many HPCC projects relating to networking is the active participation by the computer and communications industries. For example, in the Gigabit Testbed program, ARPA and NSF fund university researchers, while industry provides the experimental facilities as well as salaries for those of its employees working with the universities. Six testbeds, each with a different focus on technology and applications, have been implemented in the past three years. Another, similar HPCC university-industry collaboration is studying the science and technology of optical networks. These projects provide a good model for future collaborations and demonstrate the synergy that can result from government leadership in helping to set directions, in bringing participants together, and in providing financial support for the university components.

Important results have come about from speculative basic research funded by the federal government without industrial participation. While such research is not always a good candidate for university-industry collaboration, it is essential to ensuring that a foundation will be laid for the next generation of experimental testbeds. Moreover, a vigorous program in academic research is necessary to ensure a continued flow of new professionals into the university and industrial sector. Given the recognized importance of information infrastructure to the nation, it seems critical to ensure that an adequate number of graduates enter the field. However, there is anecdotal evidence that graduate students are choosing not to enter the field of networking, and that new PhDs are being discouraged from academic careers because it is so difficult to obtain adequate funding.[38]

The committee recognizes a number of critical objectives that can be met through a strong continued program in research and development. *The government must maintain and expand a vigorous program of research in communications, networking, information infrastructure, and basic systems research. This program should (1) include a strong experimental component, with prototype development and testbed experimentation, and (2) involve academia and industry, with specific attention to the factors that have led to the great success of the past governmental efforts such as the Internet: vision, leadership, technical strength within the government, and cooperation among the sectors of the government to achieve a common goal.*

Despite high-level administration support for the principle of technological advance, concerns expressed within the Congress and the administration raise questions about the availability of resources in general for infrastructure-related research and development and about the future of the HPCC program in particular.[39] Federal support for HPCC or its essential components relating to computing and communications technologies will help to determine how well founded are assumptions about the continued rapid advance in the high-performance computing and communications technologies that underpin the features of an NII.

A number of key areas in which specific technical innovation is required in order to accomplish the goals of the NII are discussed in Chapter 2. These key areas include the following:

- Core architectural issues for the NII, including the definition and validation of a suitable bearer service, a framework for security, and solutions to key problems in scaling, such as addressing, routing, management, heterogeneity, and mobility;

- New technologies for access circuits and related network technology, which could provide cost-effective delivery of high-peak-rate traffic to and from the end locations;
- Key middleware services, including new models for organizing and exploiting on-line information, models for managing intellectual property rights, and a framework for electronic commerce; and
- Computer and communications security.

Beyond the specific objectives identified for the NII, there are a number of areas in which advanced research may lead to fresh capabilities for the network beyond the ones envisioned today. Such capabilities will be key to the evolution of the NII into yet more powerful and useful constructs over time. In addition to a directed effort addressing the specific technical needs of the NII, a broad program of advanced network research should be maintained, as the committee has recommended in Chapter 2.[40]

CONCLUSION

The NII initiative presents exciting opportunities for the federal government to reap far greater returns from the NREN program than those experienced to date and to meet a broad range of social and economic needs. The NSF and other HPCC agencies have opportunities to lead in the development of general and flexible architectures and to experiment with their implementation. NIST and other agencies have opportunities to promote more effectively the kind of standards that will be needed to assure the broad interoperability characteristic of the Open Data Network described by this committee. Above and beyond the roles that seem obvious for individual agencies is a need for sustained leadership and effective coordination—for management in the best sense, reflecting the recognition that the federal role is one of catalyst rather than performer for most of the actions necessary to implement the NII.

NOTES

1. One informal measure of the confusion surrounding the NREN program was reflected in the committee's efforts, among its members and those from government and elsewhere who briefed it at its early meetings, to define the program. These internal discussions revealed considerable lack of certainty and agreement.

2. Committee on Physical, Mathematical, and Engineering Sciences, Federal Coordinating Council on Science, Engineering, and Technology. 1994. *High Performance Computing and Communications: Toward a National Information Infrastructure*. Office of Science and Technology Policy, Washington, D.C.

3. Note that a by-product of leadership in the development and use of infrastructural technologies should be strength in international markets. Research, development, and application are only some of the steps that can be taken toward that end. Other areas, involv-

ing the National Institute of Standards and Technology, the Departments of Commerce and Defense, and other federal agencies, include assuring that as much U.S. technology as possible can be exported for sale in foreign markets and assuring that international connectivity between the U.S. information infrastructure and counterparts overseas is maintained and advanced.

4. Until now, federal investment has not played a big role in developing information resources compared to private investment, except in scientific research and certain government information arenas.

5. CSTB held a workshop examining the spectrum of positions and options in the areas of regulation and public investment, insights from which will be captured in a forthcoming report.

6. The "Telecommunications Policy Reform" release from the Office of the Vice President, January 11, 1994, summarizes four proposals: make the preservation and advancement of "universal service" an explicit objective of the Communications Act; charge the Federal Communications Commission (FCC) and the states with continuing responsibility to review the definition of universal service to meet changing technological, economic, and societal circumstances; establish a federal-state joint board to make recommendations concerning FCC and state action on the fundamental elements of universal service; and oblige those who provide telecommunications services to contribute to the preservation and advancement of universal service. These proposals build on principles articulated by Vice President Gore in a December speech, including the following: encourage private investment, provide for and protect competition, provide open access to the network, avoid creating information "haves and have nots," and encourage flexible and responsive government action.

7. An encouraging sign comes from a survey sponsored by IBM, which found that nearly half of small business executives queried in January 1994 had heard of and were interested in using the "Information Highway." Roper Starch Worldwide Inc. 1994. "The IBM Survey on Small Business and the Information Highway," IBM, January.

8. There is now a prospect that multiple providers—at least two, a telephone and a cable company, and quite possibly wireless alternatives—may invest in local infrastructure. For example, alternative access companies were recently allowed by the FCC to provide switched access services, and most of the major multiple system operators in the cable industry are already deploying fiber in their backbone feeder routes as well as testing technologies to support two-way communications services over their existing distribution networks. It is not clear that these redundant efforts will convey proportionate benefits, especially if the cost of this competition in attractive market areas is further delay in upgrading infrastructure in more sparsely settled areas.

9. A related need is training to promote responsible use of networks and network-based resources and services.

10. For example, NSF announced in October 1993 a new program on Networking Infrastructure for Education, jointly run by the Computer and Information Science and Engineering and the Education and Human Resources Directorates. It will be linked to the NREN program, and it focuses on science and mathematics education, through anticipated testbeds, infrastructure, and tool development.

11. See, for example, "FARNET Report on the NAP Manager/RA and vBNS Provider Draft Solicitation to the National Science Foundation Division of Networking and Communications Research and Infrastructure," August 3, 1992, paper distributed by FARNET, Waltham, Mass. It presents a comprehensive overview of financial and other issues related to NREN. In particular, the report points out that "the financial implications of various strategies are not homogeneous. . ." and that considerations should be identified from the perspectives of the various entities.

Also, note that when the federal backbone efforts began there were no commercial providers capable of offering required services. Now components of the Internet, they have served as testbeds for experimentation with network services, algorithms, software, and hardware.

12. In research and education alone, a predominantly federally supported backbone and network complex is already being replaced by a broader mix of specialty backbone networks (NSI, ESnet, vBNS), commercial backbone networks (CIX, ANS CO+RE, CoREN, networks of telephone companies), and other players (regions, states, metropolitan area networks, competitive access providers, cable companies, and others).

13. Several groups have issued reports in the past year with the intention of influencing public policy and attitudes regarding the NII. See, for example, Computer Professionals for Social Responsibility (CPSR). 1994. *Serving the Community: A Public Interest Vision of the National Information Infrastructure.* CPSR, Palo Alto, Calif.; and Council on Competitiveness. 1993. *Competition Policy: Unlocking the National Information Infrastructure.* Washington, D.C., December.

At the same time, individual companies have spent considerable sums trying more directly to persuade government officials to share their views on related issues. For example, "In 1992 alone, Federal Communication Commission records show, the Bells' telephone operations poured $40 million into lobbying, a sum that the Bells' long-distance rivals claim is two to three times as much as they collectively spend." Telephone companies also maintain a presence in every major community, aiding in their interactions with state and local officials. Wartzman, Rick, and John Harwood. 1994. "For the Baby Bells, Government Lobbying Is Hardly Child's Play," *Wall Street Journal*, March 15, pp. A1 and A10.

14. Ten agencies participate in the NREN component as users; each implements its own NREN activities through normal agency structures and coordination with OMB and OSTP. Multiagency coordination on related research is achieved through the HPCCIT subcommittee, the HPCC National Coordinating Office, and the HPCCIT high-performance communications working group.

Operation of the federal components of the Internet is coordinated by the Federal Networking Council (FNC), which consists of agency representatives. The FNC and its executive committee establish direction, provide further coordination, and address technical, operational, and management issues through working groups and ad hoc task forces. The FNC has established the Federal Networking Council Advisory Committee, which consists of representatives from several sectors, including library sciences, education, computers, telecommunications, information services, and routing vendors, to assure that program goals and objectives reflect the interests of these broad sectors. Committee on Physical, Mathematical, and Engineering Sciences, Federal Coordinating Council for Science, Engineering, and Technology, Office of Science and Technology Policy. 1994. *High Performance Computing and Communications: Toward a National Information Infrastructure.* Office of Science and Technology Policy, Washington, D.C., p. 34.

15. The committee acknowledges that some public interest advocacy groups (e.g., the Coalition for Networked Information, Electronic Frontier Foundation) have begun to raise similar concerns, but no one has fully addressed the collective concerns of the research, education, and library communities, let alone those of associated public-interest communities such as health care.

16. The distinction was advanced, for example, during presentations by officials at the March 15-18, 1994, High Performance Computing and Communications Symposium organized by ARPA; it has also been posited during more informal communications between federal officials and committee members and staff in March and April 1994.

17. Building on successes in developing the HPCC and other initiatives, proposals associated with the National Performance Review include the notion of a "virtual agency" composed of relevant offices and personnel from several otherwise separate agencies.

18. An independent agency rather than an agency reporting to the administration, the FCC—assuming proposed regulatory changes are successfully enacted—will have continued involvement through its regulatory and limited standards-setting mission.

19. "Draft Report of the Federal Internetworking Requirements Panel," prepared for the National Institute of Standards and Technology, January 14, 1994.

20. Other federal agencies can also contribute to the technology development process; the group of NREN agencies has individually and collectively contributed substantially to network-related technology development. The emphasis here on ARPA and NSF derives from their more broadly cast missions as well as their histories in this area.

21. In an address to a United Nations conference on telecommunications, Vice President Gore referred to an expectation for a "planetary information network" that would be achieved without U.S. funding, through private investment. See Nash, Nathaniel C. 1994. "Gore Sees Privatization of Global Data Links," *New York Times*, March 22, p. D2.

22. A new National Research Council study, "Bits of Power," is examining some of these issues in the context of research networking.

23. X.400 and X.500, part of the OSI suite, are in some use.

24. Comments on the draft included criticisms by aerospace firms, manufacturers, and foreign governments of the prospect of OSI abandonment given investments to date in support of OSI as well as criticism that individual agency choices of protocols could be anarchic; overall, however, comments from U.S. parties supported the draft FIRP report by a factor of two to one. See Messmer, Ellen. 1994. "Critics Assail Plan to End Fed's OSI Policy," *Network World*, March 21, pp. 1 and 63; and Masud, Sam. 1994. "Agencies Question Wisdom of Opening up GOSIP," *Government Computer News*, March 21, pp. 1 and 104. The two-to-one support figure was reported by FIRP member Milo Medin at an IETF meeting in March 1994.

25. "Draft Report of the Federal Internetworking Requirements Panel," prepared for the National Institute of Standards and Technology, January 14, 1994, p. vi.

26. *Wall Street Journal*. 1994. "Blocking the Information Highway," April 8, p. A14.

27. Olsen, Florence. 1994. "Feds Define Wireless Needs As Spectrum Auction Nears," *Government Computer News*, February 21, pp. 48-49.

28. According to Scott Shenker, in a discussion of the challenge of making the "right decisions" about networking service interfaces, "Once a home consumer standard becomes widely adopted, there is tremendous consumer pressure for that standard to remain stable. . . ." See Shenker, Scott. 1993. "Service Models and Pricing Policies for an Integrated Services Internet," prepared for a conference, Public Access to the Internet, John F. Kennedy School of Government, Harvard University, May 26-27, June 8 version.

29. "Draft Report of the Federal Internetworking Requirements Panel," prepared for the National Institute of Standards and Technology, January 14, 1994. The committee recognizes that there remains a problem—the problem that motivated the FIRP—of making premature commitments to what may prove to be the wrong technologies, as some believe happened with the federal GOSIP standards.

30. However, since users connect to ESnet, NSI, and NSFNET via regional networks, network traffic may be carried by any of these backbones depending on circumstances. For example, an outgoing message from a university scientist to a NASA investigator may be carried by the NSFNET backbone while the return message may go over the NASA Science Internet (NSI) backbone. The extension of such sharing to other, commercial service providers raises questions about funds flows, the prospects for some kind of settlement arrangements (as is currently found in telephony), and implications for arrangements, such as the voucher concept discussed in Chapter 5, that direct funds to service providers.

The relationships among the three federal research backbone networks is an important consideration for the immediate and long-term future of research networking. If the NASA and DOE backbones continue in their present form after NSF withdraws from the current

NSFNET backbone, patterns and expectations of use may change. If the commercial replacements for relevant applications and/or the new vBNS become the new central transfer modes, the NASA and DOE backbones may be used less—assuming, based on current conditions, that they have slower performance. Or, the commercial charges for interregional connectivity may encourage users associated in some way with NASA or DOE to transfer their network use to these agencies' backbones despite slower transfer rates. There is some concern among researchers who use multiple federal backbones that the changes in NSFNET may result in differential charging and administrative rules plus differential connectivity in a context in which users may have relatively little input.

31. In principle, there could be considerable savings in consolidating multiple networks but a great deal of engineering would be needed, as well as determination of the extent to which facilities and clientele are duplicated.

32. A worst-case scenario envisions a power struggle destroying routing coherence, only small groups able to talk to each other, government stepping back in with a heavy hand, and a new regulatory commission or regime. There is intuitive appeal to contemplating taking a few small steps now to get it cleanly into private hands.

Note that over the last 20 years, the evolution of the Internet has been guided by a group that, while it has renamed and reorganized itself a number of times, has clearly provided the continuity and direction for the protocol development and the technology deployment. At the present time, this group is manifested as a professional organization called the Internet Society, under which is found the Internet Architecture Board and all the working groups that collectively form the Internet Engineering Task Force. This group is responsible for the development and approval of Internet standards, which are published as "RFCs" (requests for comments), as well as for starting new working groups in areas needing fresh initiatives.

33. The architecture is complemented by plans for a "metacenter" connecting the four NSF supercomputer centers, access to which would be supported by the vBNS.

34. ARPA recently funded Bell Atlantic to develop a Washington area interagency Advanced Technology Demonstration Network involving 2.4-Gbps SONET and ATM technologies supporting desktop applications using from 100 Mbps to 1 Gbps (Masud, S.A. 1993. "ARPA, Five Agencies Will Test Tomorrow's High-end Network," *Government Computer News*, November 22, pp. 1 and 52).

35. Note that K-12 schools sometimes prefer the cheaper option of buying basic infrastructure to purchasing the higher-value-added integrated service.

36. In March 1994, the Senate passed H.R. 820, enfolding S.4, included Title VI, the "Information Technology Applications Act of 1994," the origins of which date to legislation proposed by Albert Gore as a Senator in 1992. Section 102 amends the High-Performance Computing Act of 1991 provisions for the NREN program, indicating that program funds should be targeted to acquisition of commercially available communications networking services; "customized" services may be contracted for if commercial services are not available. This formulation is considered to relax the strict dichotomy between experimental and production networks advanced earlier (beginning in 1992) in the development of this legislation.

37. "Leading Telco CEOs Jointly Support Clinton-Gore Technology Initiative." 1993. Press release transmitted via electronic mail, March 26. Telecommunications executives called for "a shift of emphasis from government's direct support of networks," including ". . . a target structure comprised of separate Experimental and Production Networks."

38. National Science Foundation. 1992. *Research Priorities in Networking and Communications*, workshop report. NSF, Washington, D.C., April, p. 5. The broader problem of disincentives facing students and faculty interested in experimental computer science is discussed in: Computer Science and Telecommunications Board, National Research Council.

1993. *Academic Careers for Experimental Computer Scientists and Engineers.* National Academy Press, Washington, D.C.

39. CSTB is currently assessing the HPCC program, pursuant to a legislative request.

40. Committee members also participated in the winter 1994 NII Research Forum hosted by NIST and sponsored by several organizations, which generated a wide-ranging list of detailed recommendations for research that is believed to complement the higher-level list presented in this report.

APPENDIXES

APPENDIX A

Federal Networking: The Path to the Internet

DEVELOPMENT AND GROWTH OF THE INTERNET—A THUMBNAIL SKETCH

The Federal Networking Legacy

Federal networking activities, in the context of this report, originated with the ARPANET program, itself the seed for the larger and more amorphous Internet, which in turn has provided a foundation for the National Research and Education Network (NREN) program. Beginning in the late 1960s the U.S. Defense Department's Advanced Research Projects Agency (ARPA) funded the high-risk, high-payoff development of the ARPANET, a 56-kbps backbone network.[1] The first ARPANET node was installed at the University of California at Los Angeles in September 1969, thus launching the first packet switching network; by 1971 approximately 20 nodes had been installed, and ARPA was funding 30 different university sites as part of the ARPANET program. In the mid-1970s, the TCP/IP protocols were developed to link together different packet networks. In the late 1970s and early 1980s, research versions of local area networks and workstations were connected to the ARPANET, thus forming what is now more widely known as the Internet. At the time of the transition from the original ARPANET protocol, the Network Communication Protocol, to TCP in 1983, only a few hundred computers were on the nascent Internet, which connected only a handful of networks.[2]

Use of the ARPANET was expanded in the 1970s to the computer science research community and to segments of the science research community supported by the federal mission-oriented agencies. As a result of this expanded use and more regular operations, in July 1975 ARPA transferred ARPANET management (previously the responsibility of Bolt, Beranek, and Newman) as well as the Network Measurement Center (previously the responsibility of UCLA) to the Defense Communications Agency (DCA; now the Defense Information Systems Agency) with the expectation that direct experience with packet switching by DCA would ultimately be of wider benefit to the Department of Defense. Meanwhile, during the late 1970s and early 1980s, a number of federally supported, discipline-specific networks were established. Among these was MFEnet, which was funded by the Department of Energy (DOE) to give academic physicists doing research in nuclear fusion access to supercomputers at the Lawrence Livermore National Laboratory. MFEnet specified and implemented its own communications protocols. Two other federally supported networks were the DOE-funded HEPnet (for high-energy physics research) and the Space Physics Analysis Network (SPAN) funded by the National Aeronautics and Space Administration (NASA). SPAN and HEPnet were part of a global DECnet-based network that supported collaboration by scientists working in the space sciences and physics.

The National Science Foundation (NSF) has played a critical role in the second decade of research networking's evolution, funding research into relevant technologies as well as network deployment and use. In the period from 1980 to 1986, NSF supported the development of CSNET, a "logical network" for computer science researchers. CSNET was a network of networks, one component of which used the Internet protocols over an X.25 public data network. It also included the ARPANET and PHONENET, a telephone-based electronic mail relaying system. By 1985, CSNET had links to over 170 university, industrial, and government research organizations and numerous gateways to networks in other countries. In 1987 CSNET merged with BITNET, a network serving users from academic institutions that was initiated in 1980-1981. CSNET operations were continued under the Corporation for Research and Education Networking, whose operating costs were completely covered by member organizations' dues. Its mission apparently accomplished, CSNET service was discontinued in the fall of 1991.[3]

In 1985, NSF initiated a program of networking and computer support for centers with supercomputers to be used by researchers across the science and engineering research community. This program began with a memorandum of understanding with ARPA to allow NSF-funded supercomputer centers and selected researchers to use the ARPANET. NSF instituted the NSF Connections program in 1986 to broaden the base of network users with their own computer facilities and eventually to help universities achieve access to supercomputers (by supplying supporting hardware and/or telecommunications lines for direct, point-to-point connections); also in 1986, it launched the NSFNET network backbone program. After significant congestion was experienced in 1987, the backbone was upgraded to T1 service (1.5 Mbps) that became operational in 1988. Between 1986 and 1987, the point-to-point Connections program grew to include support for more complete networking arrangements that would collect network traffic and provide access to the limited NSFNET. That period began with funding for SURAnet and NYSERNet regional network proposals and saw over a dozen regional networks come into being (e.g., BARRnet, Midnet, PSCnet, Sesquinet, and Westnet).[4] The Connections program evolved to emphasize connecting universities and other research and educational institutions to the Internet, and under the new very high speed backbone network (vBNS) orientation (see "Continuing Evolution of Provisions," below) it is expected to support institutional needs for vBNS access (with awards based on evaluation of competing proposals).

The NSF's funding arrangements have given rise to a three-tier structure of campus (primarily universities and research institutions), regional, and backbone networks serving the research community (see Figure 1.2 in Chapter 1).[5] This conceptualization is complemented by commercial networks, international networks, and other interconnections that have resulted in a globally interconnected mesh of networks known as the Internet.[6]

NSFNET—The Research and Education Communities' Link to the Internet

The late 1980s witnessed what might be called a rationalization of federal networking activities of the previous two decades. Initial hardware, software, and reliability problems among regional networks helped to motivate the development of network operations centers; such a center became an element of the 1987 NSFNET backbone solicitation. According to one characterization, "NSFNET has a very special role in the [Internet] hierarchy: it acts as a generic transit, routing, and switching network for research and education networks."[7]

Synergistic Growth of NSFNET and the Internet

The concept of what the NSFNET could be and recognition of the problems it posed for parties operating and using it expanded steadily through the late 1980s and early 1990s. The backbone network speed was upgraded from 56 kbps to T1 and then T3; this backbone (45-Mbps) network grew to a size of 19 nodes, including 16 sponsored by NSF and 3 serving interagency communication needs. In 1989, 20 years after its birth at UCLA, the ARPANET was officially decommissioned; its descendent, the NSFNET, inherited its role as the research and education communities' backbone network.

Through the NSFNET program NSF spurred the growth of the Internet and provided a national demonstration of the feasibility of the technology and the market potential for open data networking. NSF also nurtured the broadening of the community served by the NSFNET through its requirements that the supercomputer centers and regional networks it supported develop their markets and move toward self-sufficiency; the phased withdrawal of financial support for regional networks, now linked to the termination of support for the NSFNET backbone, is the latest manifestation of NSF's promotion of self-sufficiency in the dispersed network-based activity it has fostered.

Continuing Evolution of Provisions for Research and Education Networking

As discussed in Chapter 5, the shift from direct provision of broad-based backbone network service through the NSFNET backbone to an expectation that commodity backbone and interregional networking services will be procured directly by users or their institutions is another part of a pattern of emphasis on eventual self-sufficiency. The new NSFNET, circa 1994, will involve a four-node (at least initially), very high speed backbone network service (vBNS), with network access points (NAPs) for connections among various networks serving the research and education communities and others (see Box 1.3 in Chapter 1). The NAPs themselves may eventually be turned over to participating networks for funding as well as management, much as commodity backbone transport is being handed off imminently.[8] NSF has indicated that support for connection to the vBNS is expected to be linked to proposals for expanding enhanced capabilities provided by regional networks, which will effectively serve as capillaries for an otherwise small vBNS.[9]

The National Research and Education Network Program—Expansion from the Internet Base and Earlier ARPANET Efforts

The NREN program emerged during the period from 1987 to 1989 as part of the launch of the High Performance Computing program (now known as the High Performance Computing and Communications (HPCC) initiative). Within the NREN framework, different federal agencies (notably, NSF, NASA, DOE, and ARPA) launched or expanded separate but interconnected networking efforts; these became components of the Internet.[10] These networks resulted from the early understanding by the four dominant NREN agencies of the potential value of such services to the researchers that they fund.[11]

Each agency backbone network utilizes packet switching technology developed by the ARPANET as well as the results of early Internet projects at universities and national laboratories plus complementary research in industry; many of these projects fueled commercial spin-offs and business development. The ARPA packet radio program, for example, supported development of the protocols that are the basis of the Internet; early protocol developers used the ARPANET to support their collaboration. The earliest TCP protocol work was undertaken in the mid-1970s at Stanford University, Bolt, Beranek, and Newman (BBN), and University College in London; the Internet work was thus international from the outset. The ARPANET also inspired a number of commercial data networks (such as Telenet) in the mid-1970s.

Mission agency networks serve specific communities: space and earth scientists in the case of the NASA Science Internet; scientists doing energy-related research in the case of ESnet; defense-related researchers and others in the case of the various networks of ARPA, including DARTnet and especially the Terrestrial Wideband Network TWBnet; and the broadest assortment of scientists and disciplines in the case of NSFNET. Some agencies, such as the National Institutes of Health and within it the National Library of Medicine (NLM), have launched network-based information services that depend on additional commercial networks for access.[12]

The four principal HPCC agencies also support the development of new and enhanced networking and related technologies under the NREN umbrella. ARPA, for example, has emphasized scalable architectures as well as high-speed computing and networking, exploring concepts and implementation issues with NSF through the gigabit testbed activities and now also through the optical network program.[13] DOE and NASA support development of "enabling technologies" as well as applications in support of their mission needs. For example, DOE has contributed to the CASA gigabit testbed through Los Alamos National Laboratory efforts. This work supported DOE's mission interest in high-speed networking and progress in global climate modeling, quantum chemical reaction dynamics, and three-dimensional seismic profiling.

Growth of Internet Use and Development Beyond the Research Community

Connectivity for the research community has driven the development of the Internet, but over the past five years access by the K-12 education and library communities has grown more rapidly, aided in part by outreach programs from the research community and the HPCC funding agencies. DOE, NASA, and the National Oceanic and Atmospheric Administration (NOAA), for example, have begun to package and provide access to scientific data and tools in programs intended for K-12 educational applications. NSF has also supported the broadening of Internet access through pilot projects for K-12 networking support through the Education and Human Resources Directorate and through pilot projects for library networking supported by the Department of Education's Title II-D program. Finally, HPCC agencies have also reached out to other, specialized communities through specific applications and services. The NLM's programs aimed at health care researchers and parties within the health care delivery system are an obvious and important example.

Internet product development has also grown significantly since 1980, building on federally funded research. In addition to the commercial spread of the TCP/IP protocols, a related development was the incorporation of the UNIX operating system into local area network (LAN) and high-performance workstation products, building on efforts supported by ARPA at BBN and at the University of California at Berkeley. In the 1990s, growth in commercial activity has been most evident for communications and information services, growth for which NSFNET has been a catalyst. For example, PSI was spun off from NSYERNet; ANS was an outgrowth of the Merit regional networking activities; CERFnet was launched by General Atomics, which also runs the NSF-funded San Diego Supercomputer Center.[14] Around 1990 a conscious effort was made to link commercial and nonprofit information service providers such as Dow Jones, Dialog, NLM, and CARL to the Internet.

Transition to a New Era in Networking

The latest federal initiative relating to networking is the expansion of the HPCC to include the new Information Infrastructure Technology and Applications (IITA) program. This direction was made possible by the High-Performance Computing Act of 1991 (PL 102-194), although specific funding authorization did not come until 1994. Agency program activity under IITA commenced in 1993, building on the generation of ideas relating to so-called National Challenge areas (manufacturing, education, health, and libraries). According to an official description of IITA activities, the Gigabit Testbed program will continue to foster advanced technologies and their integration in support of prototyping for national challenge efforts. Activities will fall into four broad areas: information infrastructure services; systems development and support environments; intelligent interfaces; and national challenges (apparently, specific applications).[15] At this time it is not known how—in terms of both complementarity and substitution—the growth of IITA will affect activities under the NREN umbrella.

THE INTERNET TODAY—WHAT IT IS AND HOW IT WORKS

The Internet, a global interconnection of computer networks spanning 61 countries, over 2 million host computers, and an estimated 15 million users, provides paths for communication among these computers. This interconnection permits a range of activities to be accomplished—among them exchange of electronic mail (e-mail), exchange of files, and remote login to computers—and provides access to a growing array of on-line information. Used today by many different communities in support of collaboration, cooperation, and dissemination of information, the Internet is viewed by its creators as a public resource.

The Internet is an open network. Any individual or organization is welcome to attach and become a part of the Internet. All that is required is a terminal or a computer with the correct software and the ability to pay the costs. The necessary software is now available to attach nearly all kinds of computers, including personal computers, and the options for attaching are expanding. For the most part, there are no limitations on the purposes for which the network may be used.[16]

Features and Use of the Internet

The Internet's Services

The most popular service of the Internet has been e-mail. It exists in several forms, including mail between individual users and mail to "newsgroups," which are ongoing group discussions on many different topics. Although mail and newsgroups may dominate Internet usage today, this pattern is changing with the rapid rise in applications that provide access to information (Figure A.1). Tools such as World-Wide Web (and its most popular user interface, Mosaic) are transforming the use of the Internet. Chapter 2 includes information on these and other new Internet services.

Another new service now in experimental use but not yet in production is multimedia teleconferencing. In addition, audio, video, and shared work-space tools are now emerging from the research community, and content is coming from a variety of multimedia information providers. The coming availability of such services and the expected wide deployment of the next generation of very powerful workstations will have considerable impact on the required capacity and resultant costs of the Internet. (Chapter 5 discusses in some detail the issues of costs and pricing for use of the Internet.)

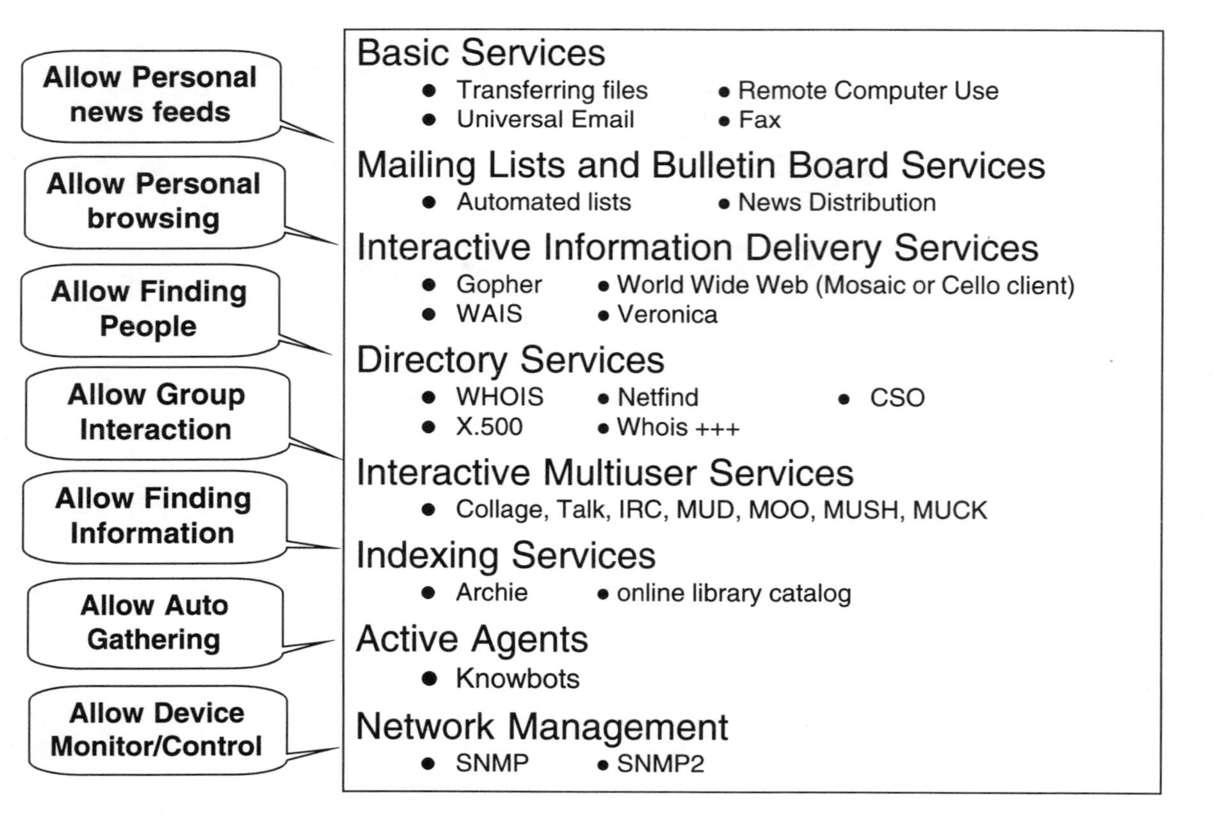

FIGURE A.1 Internet services: global information space. Courtesy of the Internet Society, Reston, Va.

How Can One Attach to the Internet?

Most users of the Internet today are affiliated with institutions. The computer they use is attached to a LAN, to which other end users are also attached; this LAN, in turn, is connected by a router to a regional or wide area network. For users located in homes or at isolated business sites, there are two ways to attach. One is to use a terminal to connect to a computer (host) that is already attached to the Internet—a very common approach to attachment today. The other is to equip the local computer with Internet software and connect not to a remote host, but to an Internet router. This mode of connection, although somewhat more complex (and more expensive), permits the local computer direct access to a fuller range of services, for example file transfer.

It is also possible to use a dial-up connection to pass packets into the Internet; two of the common protocols for this purpose are SLIP and PPP, which are often mentioned in software packages offered for personal computers. The limited bandwidth of dial-up lines may constrain the use of some Internet services such as teleconferencing, but there are today few cost-effective alternatives at higher speeds. This issue is discussed further in Chapter 5.

What Does It Mean to Be a Part of the Internet?

To connect to the Internet, an end node must run software that implements what are called protocols. The various protocols of the Internet define the packet formats, the rules for packet exchange, the higher-level services available, such as forwarding of e-mail, and so on. The Internet protocols are often known by the names of the most important two, TCP and IP. Many computers now come with the software for Internet connection built in, and such software is commercially available for essentially any other computer today. Computers in the IBM PC class, the Macintosh, and a range of UNIX machines are all very common end nodes on the Internet.

Because the Internet is a somewhat amorphous collection of networks and end nodes, the question of what it means to attach to or be a part of the Internet can cause confusion. At the technical level, an end node is a part of the Internet if (1) it implements the required protocols (the Internet protocols, including IP and TCP), (2) it has an Internet address that includes a network number known in the routing tables in the Internet routers, and (3) it has some sort of communications connection that permits it to exchange network information elements (IP packets) with the other 2 million machines on the Internet.

An alternative to connecting to the Internet at the low (packet) level is to connect at a higher (application) level. Even without packet-level connectivity, higher-level services can be interconnected through some sort of gateway. Thus, for example, interconnected mail systems permit the exchange of e-mail among a pool of users that is much larger than the group of users who have IP-connected hosts for mail exchange. Although users interconnected through a relatively high-level (application) gateway may feel that they are "on the Internet," their perspective is limited; as new services become available over the Internet, mail-only users will discover that such services are not accessible to them.

Of the large private corporate networks, some have been implemented using the Internet protocols, and some using other, vendor-specific protocols. Because of concerns about possible breaches of security, these networks are not usually connected directly to the Internet by a packet router. Instead, a higher-level service gateway is more commonly used, which does permit mail interoperation but may limit other services. Chapter 2 discusses the importance of designing a network architecture so as to provide users a range of choices in level of connectivity to the network, and it addresses issues of network security and its relation to packet routers and higher-level service gateways.

Internal Components of the Internet

The typical user may be satisfied to know that the Internet is a system capable of moving data from one place to another as requested by the attached computers. But within the Internet is the wide range of technologies from which it is built.

The core technical concept of the Internet is packet switching, which was proposed about 30 years ago as a very efficient way of sharing very expensive long-distance telecommunications circuits. A packet (a small amount of data with a destination address on the front) can be put in line with packets from other hosts and sent in turn down a link. This very fine grained sharing allows the cost of the link to be divided among many users.

Internally, the Internet is composed of networks, almost 20,000 of them. A network is a logical entity that may in turn be composed of a number of physical network elements called subnetworks. Some of these networks are collections of LANs at campus locations, for example, Ethernets and Token Rings; some are long-distance networks, such as the NSFNET backbone network; and some are regional or state-level networks. Each network has a unique network number that is part of the address of the end nodes, or computers, attached to each net. Addresses are organized in a nested manner and are represented by four elements, which are written as four numbers separated by dots. Sometimes more than one element is used to name a network or a subnet—if we used just one field to name networks, we could name only 256 nets, because each field holds only 256 values. The address of a particular machine at the Massachusetts Institute of Technology, the one on which much of this appendix was written, is 18.23.0.108. This means that the machine is host 108 on subnetwork 23.0 on network 18, with network 18 covering all of MIT and subnet 23.0 covering the building in which the end node (this author's computer) is located.

The device that connects networks together and passes packets among them is called a router. Internet routers are now commercial products, and many thousands of routers currently exist within the Internet. As the name indicates, routers deal with routing, addressing, management of telecommunications links, and other issues of operation. In particular, routers maintain routing tables, which indicate for each of the known network numbers how to reach that net. At the end of March 1994 there were 19,373 networks active in the Internet.[17] With the growth of the Internet, routing tables have become larger, and maintaining them in a consistent state is now becoming more challenging. Routing tables are maintained through a combination of on-line automatic message exchange and manual control. In early 1994 NSF announced an award for an Internet routing arbiter project intended to facilitate the logical interconnection of the networks attached to the Internet.[18]

The design of the telecommunications technology out of which the Internet is built has matured over the last decade. When the Internet was first conceived, the only wide area technology available consisted of simple point-to-point telecommunications links operating at various speeds. Now, a range of very sophisticated technologies exists for passing packets between routers, including Frame Relay, SMDS, asynchronous transfer mode, and so on. However, this increase in sophistication in the internal components of the Internet has not changed its basic nature as perceived by the end user, except that the new technologies are now facilitating provision of a broader range of services, such as real-time video, and are permitting broader and less expensive deployment of services.

National and Global Interconnection of Internet Users

No single organization owns or operates the Internet. Instead, several thousand organizations separately operate and administer their individual parts, which combine to form the total Internet (Figure A.2).[19]

Most countries have some sort of backbone or long-distance network

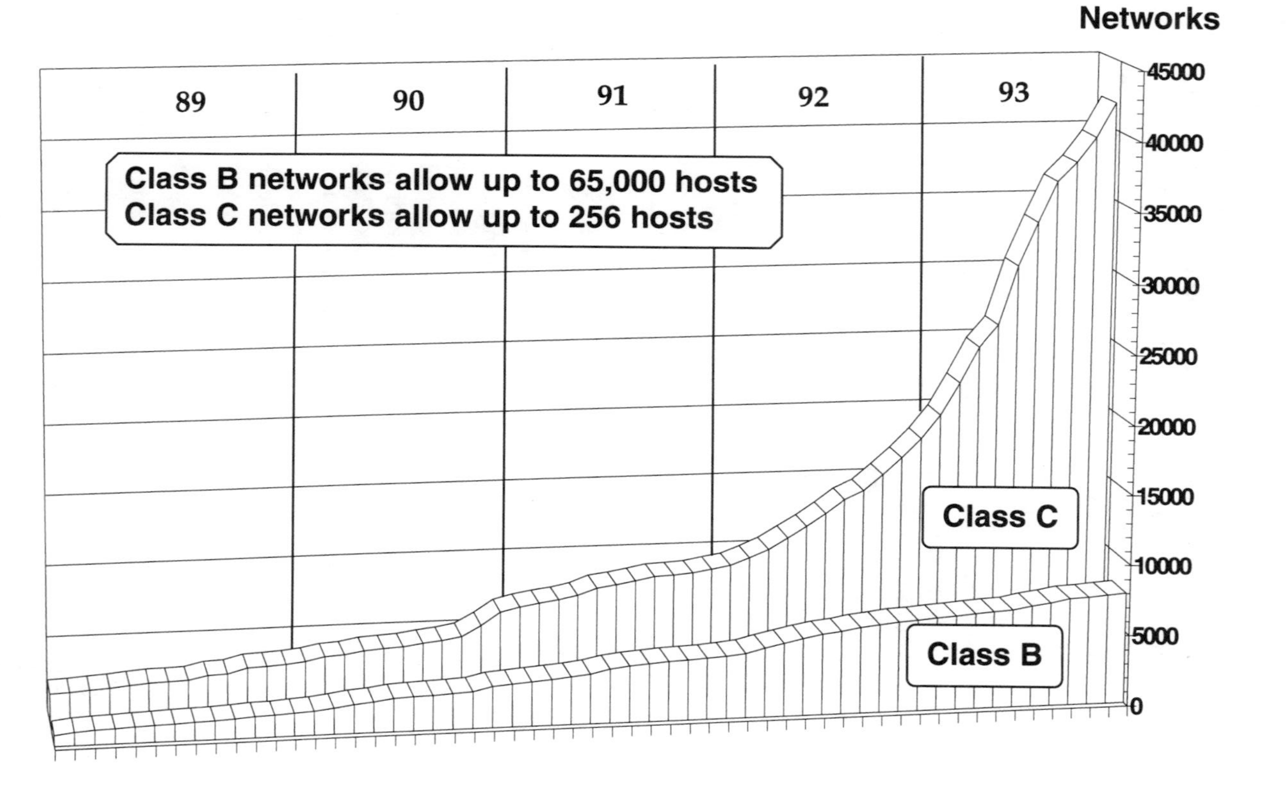

FIGURE A.2 Growth of registered IP networks, 1989 to 1993. Graph courtesy of the Internet Society, Reston, Va.

spanning the country. The United States has a number of such "nets," some provided by the government and some built for commercial use. Attached to these backbone networks are more localized networks that serve regions, states, or cities. Finally, there are the networks that cover an institution—a university, corporation, or similar facility—and there are networks that permit individuals to connect from homes and businesses.

Federal agencies have addressed the problem of interaction and interconnection within the United States through the concept of exchange points, first implemented as the Federal Internet Exchanges (FIXes; including FIX-West at NASA Ames Research Center and FIX-East at the University of Maryland, College Park). The multiple paths between the FIXes are individually subject to each agency's policy-based constraints on usage. The Commercial Internet Exchange (CIX) added a third, similar exchange point for commercial Internet access providers, and the network access points outlined in the new NSF network program solicitation (see Box 1.3 in Chapter 1) will further generalize the implementation of exchange points. Experience with and expectations for these network exchanges are motivating further work in the area of routing funded by NSF as part of its networking program.[20]

Access to and interconnection among the pieces of the Internet is offered through common carriers, leased-line access providers, and dial-up access providers, many of which are interconnected. For example, Metropolitan Fiber Systems operates fiber links between Falls Church, Virginia, and College Park, Maryland, that are known as Metropolitan Area Ethernet or MAE-East and that directly link Sprint, PSI, UUNET, NSFNET, and the main data line to Europe. Sprint leases circuits to UUNET, America Online, Digital Express, and other providers of Internet access; MCI has an equity interest in ANS and leases lines to Prodigy. Dial-up access is available from providers that range from residence-based small businesses with a single computer, modem, and telephone line to corporate communications concerns (e.g., GE Information Services or America Online).[21] Provision of access to the Internet can be and has been launched as a very small business as well as a big business; one entrepreneur reported that his total hardware cost for starting a 16-telephone-line service was under $15,000.[22]

The Internet's very decentralized administration succeeds because it is governed by a set of defining conventions (the Internet protocols and the agreed-to operating rules) that specify what each operator must do to be an effective part of the Internet. These rules define the formats of messages, how routing is done, conventions for managing the network elements, and so on.

Oversight of the Internet

Internet Society

The Internet's very decentralized organization is, of necessity, balanced by mechanisms for coordination and oversight of key issues and concerns. Of the several organizations that play a coordinating role, perhaps the most important is the Internet Society.[23] The international Internet Society provides a means for global cooperation and coordination for the Internet and its internetworking technologies and applications. Its members reflect the breadth of the entire Internet community and include individuals, corporations, nonprofit organizations, and government agencies.

The Internet Society's principal purpose is to maintain and extend the development and availability of the Internet and its associated technologies and applications—both as an end in itself and as a means of enabling organizations, professions, and individuals worldwide to more effectively collaborate, cooperate, and innovate in their respective fields and interests. Its specific goals and purposes include the following:

- Development, maintenance, evolution, and dissemination of standards for the Internet and its internetworking technologies and applications. Internet standards are formulated by the Internet Engineering Task Force, an open-membership body that currently operates under the auspices of the Internet Society;
- Growth and evolution of the Internet architecture. The Internet Architecture Board (IAB), which offers architectural guidance for the Internet, sits under the Internet Society;
- Maintenance and evolution of effective administrative processes needed for operation of the global Internet and for internetworking. For example, the Internet Assigned Number Authority, which is responsible for the allocation of network numbers for the Internet, is a function of the IAB;
- Education and research related to the Internet and internetworking, and the collection and dissemination of relevant information; and
- Harmonization of actions and activities at international levels to facilitate the development and availability of the Internet. The Internet Society is currently establishing liaison relationships with such organizations as the International Organization for Standardization and the International Telecommunications Union.

U.S. Federal Government

As outlined above, the U.S. government historically has played a very important role in the overall coordination of the Internet, but this role is diminishing with the Internet's increasing commercialization and internationalization. However, the government continues to provide some guidance in key areas, as in the recent award by NSF to provide coordination of Internet routing. In addition coordination among federal agency networking activities is provided through the Federal Networking Council (FNC; which was established in 1990 with links to the Office of Science and Technology Policy (OSTP) overall, the High Performance Computing and Communications Information Technology effort under OSTP, and the Office of Management and Budget) and the FNC Advisory Committee, which consists of representatives from the nongovernmental constituencies served by federal networks. The FNC oversees various working groups (e.g., the Engineering and Operations Working Group concerned with network technology design and implementation, and the Federal Engineering and Planning Group that conducts technical analysis and operational coordination of the federal agency networks).[24]

Paying for Use of the Internet

In the early days of the Internet, most of the funding for the facilities and also for network use came from the U.S. government, as discussed in Chapter 5. Briefly, most of the costs today are covered by payment from the various users, both institutional and individual, that attach to and are a part of the Internet. A typical payment scheme is a monthly access fee, although other patterns of payment can be found. For an institution, the costs of installing, operating, and supporting the local networks and related infrastructure at the campus are usually much greater than the access fees paid to the regional or long-distance Internet provider.

NOTES

1. Kleinrock, Leonard. 1976. "The Arpanet." Pp. 304-314 in *Queueing Systems, Volume 2: Computer Applications.* Section 5.4. Wiley Interscience, New York.

2. Cerf, Vinton. 1993. "How the Internet Came to Be." In *The Online User's Encyclopedia* by Bernard Aboba. Addison-Wesley, New York, November.

3. Cerf, Vinton. n.d. "A Brief History of the Internet and Related Networks," distributed electronically.

4. To facilitate communication and coordination among regional networks and coordination between those networks and NSF, the Federation of Academic Research Networks (FARNET) was founded independently in 1987.

5. Mandelbaum, Richard, and Paulette A. Mandelbaum. 1992. "The Strategic Future of the Mid-Level Networks." Pp. 59-118 in *Building Information Infrastructure.* Harvard University Press, Cambridge, Mass.

6. Aiken, Robert, Hans-Werner Braun, and Peter Ford. 1992. "NSF Implementation

Plan for the Interim NREN," GA-A21174, GA Project 3900, San Diego Supercomputer Center, draft, May.

7. Aiken et al., 1992, "NSF Implementation Plan for the Interim NREN," p. 8.

8. Aiken et al., 1992, "NSF Implementation Plan for the Interim NREN."

9. Stephen Wolff, National Science Foundation, remarks at National Net '94 conference, Washington, D.C., April 7.

10. "A key activity of the NREN program is to enhance the interconnection technologies and strategies for both federal and non-federal networks, without interfering with the autonomous management of each component network." Aiken et al., 1992, "NSF Implementation Plan for the Interim NREN," p. 3.

11. Computer Science and Technology Board (CTSB), National Research Council. 1988. *Toward a National Research Network.* National Academy Press, Washington, D.C. (The Computer Science and Technology Board became the Computer Science and Telecommunications Board in 1990.)

12. See agency matrix in Committee on Physical, Mathematical, and Engineering Sciences, Federal Coordinating Council for Science, Engineering, and Technology, Office of Science and Technology Policy. 1994. *High Performance Computing and Communications: Toward a National Information Infrastructure.* Office of Science and Technology Policy, Washington, D.C., pp. 26-27.

13. The Gigabit Testbed program is funded by NSF and ARPA and involves industry contributions and participation as well as participation by universities and government laboratories. See Committee on Physical, Mathematical, and Engineering Sciences, Federal Coordinating Council for Science, Engineering, and Technology, Office of Science and Technology Policy, 1994, *High Performance Computing and Communications: Toward a National Information Infrastructure.*

14. Cerf, 1993, "How the Internet Came to Be."

15. National Coordination Office for HPCC. 1994. *Information Infrastructure Technology and Applications.* Office of Science and Technology Policy, Washington, D.C., February.

16. For those parts of the Internet installed by the NSF in support of scholarly and academic activities, there are policy restrictions on the uses of the network, but these restrictions are expected to vanish from the core of the network with the transition to commercial provision of Internet service.

17. Personal communication, Elise Gerich, Merit Inc., April 1994.

18. "[The routing arbiter's] role will be to promote Internet routing and stability, establish network topology and policy databases, develop procedures to resolve problems between network entities, develop advanced routing technologies, provide simplified routing strategies, and promote distributed operation and management of the Internet." Part of the challenge is to develop new technologies, including high-performance, workstation-based route servers. See University of Southern California. 1994. "NSF Announces Major Network Awards: California, New York, and Michigan Groups Win." Press Release, February 15, distributed electronically.

19. No firm number exists for the number of organizations involved; the range is perhaps from a lower bound of 2,000 to 3,000 to an upper bound of 10,000 to 12,000. Individuals contacted by the committee remarked on how interesting the question was—it is a measure of the Internet's true decentralization that we cannot count or even identity all the Internet players.

20. Aiken et al., 1992, "NSF Implementation Plan for the Interim NREN"; and Huston, Geoff. 1993. "Connectivity Within the Internet—A Commentary." Australian Academic and Research Network, Canberra, Australia.

21. Stewart, Thomas A. 1994. "The Netplex: It's a New Silicon Valley," *Fortune,* March 7, pp. 98-104.

22. Lewis, Peter H. 1994. "A Traffic Jam on the Data Highway," *New York Times*, February 2, pp. D1 and D5.
23. Material on the Internet Society was adapted from an electronic Frequently Asked Questions document, "What Is the Internet Society?" dated March 5, 1994.
24. Aiken et al., 1992, "NSF Implementation Plan for the Interim NREN."

APPENDIX
B
Sample Principle Sets

NII Principles and Objectives

- Promote private-sector investment through appropriate tax and regulatory policies.
- Extend the "universal service" concept to ensure that information resources are available to all at affordable prices.
- Act as catalyst to promote technological innovation and new applications.
- Promote seamless, interactive, user-driven operation of the NII.
- Ensure information security and network reliability.
- Improve management of the radio frequency spectrum, an increasingly critical source.
- Protect intellectual property rights.
- Coordinate with other levels of government and with other nations.
- Provide access to government information and improve government procurement.

SOURCE: Information Infrastructure Task Force (NTIA). 1993. *The National Information Infrastructure: Agenda for Action.* NTIA, U.S. Department of Commerce, September 15.

Seven Principles from the Telecommunications Policy Roundtable

- *Universal access* — In our information age, everyone has a right to affordable news, education and government information. Information that is essential to the functioning of citizens in a democracy should be free.
- *Freedom to communicate* — Information is a two-way street. The design of the new networks should aid two-way audio and video communication from anyone to any individual, group or network.
- *Vital civic sector* — The new networks should allow all groups and individuals to freely express their ideas and opinions. The new networks should include a way for us to build communities.
- *Diverse and competitive marketplace* — No one should ever control both the wire or wires into our home and the content of the programs that go over those wires.
- *Equitable workplace* — Workers must be valued and protected in the new electronic workplace. Nondiscriminatory practices must form the core of the new information marketplace.
- *Privacy protection* — Privacy should be carefully protected and extended.
- *Democratic policy making* — Every American deserves to be heard on this complex set of issues.

SOURCE: Telecommunications Policy Roundtable. 1993. *New Coalition Unveils Public Interest Blueprint for America's 21st Century Telecommunications Highway.* Center for Media Education, Washington, D.C., via electronic mail news release October 26.

Electronic Frontier Foundation on Principles

- *Diversity of Information Sources:* Promote a fully interactive infrastructure in which the First Amendment flourishes, allowing the greatest possible diversity of view points;
- *Universal Service:* Ensure a minimum level of affordable information and communication service for all Americans;
- *Free Speech and Common Carriage:* Guarantee infrastructure access regardless of the content of the message that the user is sending;
- *Privacy:* Protect the security and privacy of all communications carried over the infrastructure, and safeguard the Fourth and Fifth Amendment rights of all who use the information infrastructure;
- *Development of Public Interest Applications and Services:* Ensure that public interest applications and services which are not produced by the commercial market are widely available and affordable.

SOURCE: Electronic Frontier Foundation. 1993. *New EFF Infrastructure Policy Statement: The Open Platform Campaign.* Electronic Frontier Foundation, Cambridge, Mass., November 3, via electronic mail from Daniel J. Weitzner, Senior Staff Counsel.

Fundamental Principles from the Computer Professionals for Social Responsibility

- *Universal access.* All people should have affordable access to the information infrastructure.

- *Freedom to communicate.* The information infrastructure should enable all people to effectively exercise their fundamental right to communicate.

- *Vital civic sector.* The information infrastructure must have a vital civic sector at its core.

- *Diverse and competitive marketplace.* The information infrastructure should ensure competition among ideas and information providers.

- *Equitable workplace.* New technologies should be used to enhance the quality of work and to promote equity in the workplace.

- *Privacy.* Privacy should be carefully protected and extended.

- *Democratic policy-making.* The public should be fully involved in policy-making for the information infrastructure.

- *Functional integrity.* The functions provided by the NII must be powerful, versatile, well-documented, stable, reliable, and extensible.

SOURCE: Computer Professionals for Social Responsibility (CPSR). 1993. *Serving the Community: A Public Interest Vision of the National Information Infrastructure.* CPSR, Palo Alto, Calif.

Draft: April 21, 1994

Principles for Providing and Using Personal Information

Preamble

The United States is committed to building a National Information Infrastructure (NII) to meet the information needs of its citizens. This infrastructure, essentially created by advances in technology, is expanding the level of interactivity, enhancing communication, and allowing easier access to services. As a result, many more users are discovering new, previously unimagined uses for personal information. In this environment, we are challenged to develop new principles to guide participants in the NII in the fair use of personal information.

Traditional fair information practices, developed in the age of paper records, must be adapted to this new environment where information and communications are sent and received over networks on which users have very different capabilities, objectives and perspectives. Specifically, new principles must acknowledge that all members of our society (government, industry, and individual citizens) share responsibility for ensuring the fair treatment of individuals in the use of personal information, whether in paper or electronic form. Moreover, the principles should recognize that the interactive nature of the NII will empower individuals to participate in protecting information about themselves. The new principles should also make it clear that this is an *active* responsibility requiring openness about the process, a commitment to fairness and accountability, and continued attention to security. Finally, principles must recognize the need to educate all participants about the new information infrastructure and how it will affect their lives.

These "Principles for Providing and Using Personal Information" recognize the changing roles of government and industry in information collection and use. Thus they are intended to be equally applicable to public and private entities that collect and use personal information. However, these Principles are not intended to address all information uses and protection concerns for each segment of the economy or function of government. Rather, they should provide the framework from which specialized principles can be developed.

I. General Principles for the National Information Infrastructure

A. Information Privacy Principle

1. Individuals are entitled to a reasonable expectation of information privacy.

B. Information Integrity Principles

Participants in the NII rely upon the integrity of the information it contains. It is therefore the responsibility of all participants to ensure that integrity. In particular, participants in the NII should, to the extent reasonable:

1. Ensure that information is secure, using whatever means are appropriate;
2. Ensure that information is accurate, timely, complete, and relevant for the purpose for which it is given.

II. Principle for Information Collectors (i.e. entities that collect personal information directly from the individual)

A. Collection Principle

Before individuals make a decision to provide personal information, they need to know how it is intended to be used, how it will be protected, and what will happen if they provide or withhold the information. Therefore, collectors of this information should:

1. Tell the individual why they are collecting the information, what they expect it will be used for, what steps they will take to protect its confidentiality and integrity, the consequences of providing or withholding information, and any rights of redress.

III. Principles for Information Users (i.e. Information Collectors and entities that obtain, process, send or store personal information)

A. Acquisition and Use Principles

Users of personal information must recognize and respect the stake individuals have in the use of personal information. Therefore, users of personal information should:

1. Assess the impact on personal privacy of current or planned activities before obtaining or using personal information;
2. Obtain and keep only information that could reasonably be expected to support current or planned activities and use the information only for those or compatible purposes;
3. Assure that personal information is as accurate, timely, complete and relevant as necessary for the intended use.

continued on next page

Principles for Providing and Using Personal Information—*continued*

B. Protection Principle

Users of personal information must take reasonable steps to prevent the information they have from being disclosed or altered improperly. Such users should:

1. Use appropriate managerial and technical controls to protect the confidentiality and integrity of personal information.

C. Education Principle

The full effect of the NII on both data use and personal privacy is not readily apparent, and individuals may not recognize how their lives can be affected by networked information. Therefore, information users should:

1. Educate themselves, their employees, and the public about how personal information is obtained, sent, stored, and protected, and how these activities affect others.

D. Fairness Principles

Because information is used to make decisions that affect individuals, those decisions should be fair. Information users should, as appropriate:

1. Provide individuals a reasonable means to obtain, review, and correct their own information;
2. Inform individuals about any final actions taken against them and provide individuals with means to redress harm resulting from improper use of personal information;
3. Allow individuals to limit the use of their personal information if the intended use is incompatible with the original purpose for which it was collected, unless that use is authorized by law.

IV. Principles for Individuals who Provide Personal Information

A. Awareness Principles

While information collectors have a responsibility to tell individuals why they want information about them, individuals also have a responsibility to understand the consequences of providing personal information to others. Therefore, individuals should obtain adequate, relevant information about:

1. Planned primary and secondary uses of the information;

2. Any efforts that will be made to protect the confidentiality and integrity of the information;
3. Consequences for the individual of providing or withholding information;
4. Any rights of redress the individual has if harmed by improper use of the information.

B. Redress Principles

Individuals should be protected from harm resulting from inaccurate or improperly used personal information. Therefore, individuals should, as appropriate:

1. Be given means to obtain their information and be provided the opportunity to correct inaccurate information that could harm them;
2. Be informed of any final actions taken against them and what information was used as a basis for the decision;
3. Have a means of redress if harmed by an improper use of their personal information.

SOURCE: National Information Infrastructure Task Force, Information Policy Committee, Working Group on Privacy in the NII. 1994. "Request for Comments on the Draft *Principles for Providing and Using Personal Information and Their Commentary.*" Washington, D.C., May 4. The draft *Principles for Providing and Using Personal Information* and the associated *Commentary* are the first work product of the Information Infrastructure Task Force's Work Group on Privacy. They are intended to update the *Code of Fair Information Practices* that was developed in the early 1970s. While many of the Code's principles are still valid, the Code itself was developed in an era when paper records were the norm.

APPENDIX

C

User Support Services

The increasing dependence of multiple communities of users on networks requires that the Internet have the following characteristics: it must be accessible at a low cost, be user friendly, and have a capacity and features that enable effective and meaningful access to needed resources. Also needed (at a minimum) is the following support service infrastructure:[1]

- Education and training, as well as programs and services, to assist users in utilizing the network;
- Outreach services to identify new communities of users and their distinct information needs; and
- Coordination between network providers and service organizations for an integrated approach to supplying user support services and access.

The most basic of the needed services is a "white pages" capability that will permit users to find each other on the network. The government-funded Network Information Center has traditionally operated a user naming service. However, a centralized approach of the kind practiced in the past is no longer appropriate, because the network has grown, and a centralized approach does not scale well.

At this time, the government should support the planning and deployment of a distributed user naming service. It could support the Internet Engineering Task Force (IETF), in conjunction with other standards-oriented groups, in the selection of a set of protocols for this purpose. In addition, the government (perhaps through the Advanced Research Projects Agency) should fund the development of a software package that implements these protocols and is made freely available, as TCP itself was distributed initially. Prior attempts to accomplish this goal have failed, due to a lack of focus and priority. The need for a distributed user naming service is even more pressing now.

Once such a framework is in place, companies, educational institutions, professional organizations, and other such groups should begin to assemble directories of their members, which can then be incorporated into the eventual distributed naming service.[2] Although the overall framework must be centrally developed (for example, by the IETF with the support of the federal government), the data themselves should be collected and managed in a decentralized manner "close to the source," to ensure their accuracy and to allow for local access controls.

The need for user support has expanded as the offerings of the Internet have become more diverse and complex and as the user population has grown. The Internet Society through the IETF has been improving user support through the development of a series of documents that help new users to join the Internet. For this kind of effort a centralized approach works well, and it should continue.

Private enterprise and private-sector institutions have major roles to play in providing direct support to users. Answering questions, resolving service problems, and facilitating attachment are necessary services that will require substantial effort. The data network is not like a telephone. It is a complex and evolving technology, and it sometimes displays inexplicable and undesirable behavior. Assistance for users is thus critical.

Sellers of individual products today provide good user support, but they cannot be expected to solve problems related to overall system integration. Today, institutional network users are supported through a growing body of for-profit system integrators and resellers, who provide support as their major product. The major problem in this area today is the lack of a mechanism for locating and evaluating these providers. In other words, we need a marketplace. This, of course, is what the information network is intended to provide.

Individual users cannot hire a system integrator to solve their specific problems. Since most users attach to the network because of some specific professional or personal interest, such as work-related activities or a hobby, one approach to providing networking support for (some) individuals would be for special-interest groups, for example, professional societies, to contract to provide network support to their members.

Existing network providers today offer support as an important part of their overall network service. As for-profit providers move to take over the actual operation of trunks and switches, these existing network providers can continue to provide service to their clients. Such an approach would separate the business of network operations, which is becoming more clear-cut in its nature, from the still rather evolving task of effectively helping users.

Customer service is a significant source of revenue for providers of computer products. The problems of supporting users of large-scale networks are more complex, and resolving these problems should present a significant business opportunity. What is important is that users clearly understand the cost and value of this service. Today, for example, cost competition for personal computers has eliminated any cost margin for support, and many users are first pleased at the very low cost of computers, and then frustrated when they cannot get parts to work. From the very beginning, network providers should be encouraged to price service and basic network access so that the cost of each is apparent, and so that a realistic level of support can be made available.

NOTES

1. In the interest of assigning responsibility for such services, one suggestion made elsewhere was to involve the IITF in service delivery. "Draft Report of the Federal Internetworking Requirements Panel," prepared for the National Institute of Standards and Technology, January 14, 1994.

2. The harvesting of names and electronic mail addresses is becoming easier through technology, although there are drawbacks associated with these practices.

APPENDIX

D

State and Regional Networks

State networks provide local connectivity and access to wider-area services for schools, higher education, and research institutions. State networks have been wholly oriented to operations; states have procured networks in much the way that large corporations have, rather than to demonstrate or advance technology, as the federal government has often sought. Given their strong links to and interests in many segments of the public interest domain—including a variety of groups relatively new to network-based services, such as hospitals and other components of health care delivery, public libraries, and state and local government entities—states have important roles to play in the current and future development of information infrastructure.

Although virtually all states are at least planning networking strategies, implementation has been uneven, with some states manifesting more coordination and integration than others. States vary considerably in terms of size, resources to support networks, geography, and the economics of networking (e.g., networking is much more costly in Montana than in Massachusetts). States also differ in their approaches to adopting network architecture and technology, raising questions about prospects for interconnection. A major factor in developments and prospects at the state level is each state's Public Utilities Commission, which regulates intrastate telecommunications offerings and pricing.

One state that has taken a lead in networking is North Carolina, which recently announced a new state networking program to which it plans to shift the $35 million per year it spends on networking services.[1] Already, it has linked community colleges to an X.25 service and universities to the T3 CONCERT network.

In California the development of networking for the research, education, and library communities has evolved in a loosely coordinated fashion. Five major networks form the nucleus of a statewide education network. The two state university systems, the University of California (UC) and the California State University (CSU), each have networks. UCNET interconnects the nine UC campuses and CSUnet the 20 CSU campuses. In addition, there are three regional networks, BARRnet serving northern California, CERFnet serving southern California, and LosNettos serving the Los Angeles basin. These three regional networks were initially designed to serve the research community but have branched out. All five networks are connected to the Internet and also to each other. A majority of the 107 community colleges have connected to the Internet via one of these five networks. Only a fraction of the several thousand K-12 schools have been connected to the Internet, most through CSUnet. The State Library system has called for a statewide network to support multitype library cooperation and coordinated planning studies. Efforts are now under way to bring that planning endeavor into an overall strategy to develop the Golden State Education Network as a "network of networks." In mid-1993, Pacific Bell announced plans for advanced communications networks linking up to 80 educational, medical, and high-technology industrial organizations each in the San Francisco and Los Angeles areas. Known as CalREN, the program involves development and pilot-testing of innovative applications over at least a two-year period.[2]

Regional (or mid-level) networks are one response to variations in support for networking among states, although they are not necessarily a substitute for state networking. They provide economies of scale for users across a relatively wide area. Most of these organizations purchase basic connectivity from existing telecommunications providers and provide the packet switched overlay as well as user support.

Regional networks, especially those catalyzed by National Science Foundation investments, service the research and library communities and increasingly the education community (K-12 and higher education); some have a considerable base of commercial clients as well. Anecdotal evidence compiled informally by FARNET and NSF suggest that about 80 percent of network traffic is intraregional and about 20 percent is between regions. Each regional network serves about 100 to 200 client sites (i.e., research and education institutions).[3] As a group, their financial health and viability vary.[4]

With the shift toward commercial packet-switched networks plus diminishing federal support, these regional providers will have to evolve to fill new roles—it remains to be seen whether regional providers as they are currently known are a temporary phenomenon. In some instances, they may become competitive providers of network service. Toward that end, eight regional networks, organized as the Corporation for Regional and Enterprise Networking (CoREN), to provide voice, data, and video services, including TCP/IP networking, between regions and nationwide.[5]

Alternatively, regional networks may evolve into or effectively reappear as collective purchasing groups, which attempt to control cost and unify service through coordinated procurement of service. This could be a means for indirect subsidy of deployment, although some hard-to-serve research and education entities might still be unable to afford to participate. A third potential role is as a user service support organization. Training is a key issue, especially for less-sophisticated potential users. Finally, another possibility is that commercial providers might buy up today's regionals if they continue to grow.

Regional networks, as currently cast (i.e., as dominant providers of regional data networking serving the research and educational communities within their regions and providing access to federal backbones), raise questions about the ease with which other networks can interconnect and thereby gain access to the regional network clientele, notably the university community. The combination of the current sheltered status of the regional networks and the formation of CoREN raises the prospect of the regionals achieving a three- to four-year head start over other, commercial providers on developing relationships with universities for which they could provide both intra- and inter-regional connectivity. Such a head start could result in substantial barriers to the entry of new commercial providers after federal support has been phased out, unless some form of broader interconnection is required now. This observation assumes that institutions have high costs for switching among network and service providers, that they lack motivation for seeking diversity in providers, and that the telecommunications companies that provide the underlying facilities for regionals do not themselves become major players, exploiting their existing relationships with universities. The rapid growth of commercial Internet access providers suggests that these problems will not be significant.

To promote regional networking infrastructure that is of value to academic and other users and that supports competition, there is a choice: (1) Establish regional regulated monopolies that cross-subsidize within the region (charging higher prices to commercial users, and lower prices to academic users), or (2) encourage competition, such as that provided by Sprint and MCI for AT&T in long-haul telephony, with only the government providing subsidies to research, education, and other public-interest communities (e.g., hospitals). Competition will lead to reductions in commercial rates; noncommercial rates will fall if there is an injection (subsidy) of government money. Which approach is best depends on an assessment of the extent to which the market may fail to meet the needs of noncommercial users and thus compromise societal interests—a central concern with respect to the future of research and education networking. Given the overall movement toward less regulation in communications, it may be desirable to try the competitive approach for a specified period, say, five years, and to carefully monitor impacts on the research and education communities.

NOTES

1. McFadden, Kay. 1993. "High-tech Communications 'Highway' in Works," *The News & Observer*, May 4, p. 1A, 6A. The article notes that, "Under the proposal, North Carolina's information highway will be built, maintained and owned by the three participating phone companies. Although the government will set priorities on the uses of the highway, it is clear that the companies will reap two benefits. They would get the state as a client, and they would be able to link up to each other without undergoing as many arduous regulatory-clearing procedures . . . . As for the state's costs, those will come at the "user end"—that is, the classrooms, hospitals, jails and other facilities that have to buy television, computer, and telephone equipment. In all likelihood, such spending will fall under individual departmental budgets" (p. 6A).

2. Pacific Bell press release, April 2, 1993, excerpts distributed electronically.

3. Information presented to the committee by Stephen Wolff, NSF, and Laura Breeden, then executive director, FARNET.

4. Mandelbaum, Richard, and Paulette A. Mandelbaum. 1992. "The Strategic Future of the Mid-level Networks." Pp. 59-118 in *Building Information Infrastructure*. Harvard University Press, Cambridge, Mass.

5. CICnet. 1993. "CoREN Selects MCI to Join Forces on National Information Infrastructure Initiatives," press release from CICnet distributed via electronic mail.

APPENDIX E

International Issues

Because the terms "NREN" and "NII" contain the word "national," it is easy to assume that problems of developing, implementing, and using information infrastructure are largely domestic. But this is misleading, because both the markets served by U.S-owned companies and the endeavors of the research and education communities are increasingly international. Thus, NREN and NII services will operate in a context of international connectivity; U.S. entities do and will use U.S. infrastructure to communicate with parties overseas, and foreign entities will use their local infrastructure to communicate with parties in the United States.

Research networks have been international since the mid-1970s, when sites in the United Kingdom and Norway connected to SATNET, and thereby to the ARPANET. In the early 1980s, CSNET, BITNET, and UUCP all developed gateways to networks in other countries. For example, by 1984, CSNET was operating electronic mail gateways between the United States and Korea, Israel, Japan, France, Germany, Australia, and Scandinavia. In the same time frame, BITNET became international as it spread to Europe, via the European Academic Research Network (EARN). Similarly, the UUCP network developed a gateway to the European UNIX networks via Amsterdam, and the U.S. agency networks, SPAN and HEPnet, were linked to communities of interest in other parts of the world. By the mid-1980s, when the first NSFNET backbone was being discussed, electronic mail gateways already connected the various U.S. networks to a robust and growing global networking infrastructure.

An early national network project outside the United States was JANET (Joint Academic Network) in the United Kingdom. Later, national network projects were initiated in large numbers of countries on every continent. Examples include DFN (Deutsche Forschungsnetz) in Germany, UNINET in Norway, SDN in Korea, and JUNET in Japan. International collaborations included NORDUNET in the Nordic countries, EARN and EUNET in Europe, and PACOM in the Pacific Rim. The various national network projects have differed with respect to technology, financing, implementation details, and the extent of government participation and control. However, all are intended to support enhanced communication within their local academic and research communities as well as communication with colleagues in other countries.

With the advent of the NSFNET backbone, connections between the United States and other parts of the world were expanded and an attempt was made to better coordinate routing and addressing. This led to formation of the Coordinating Committee for Intercontinental Research Networks (CCIRN) and the Internet Engineering Planning Group (IEPG). The Internet Engineering Task Force (IETF), the organization that sets standards for the Internet, has also developed an international character in recent years. This emphasis is reinforced by its recent location under the aegis of the Internet Society, an expressly international organization.

U.S. government agencies have played an active role in helping to internationalize the global Internet. For a number of years, the National Science Foundation (NSF) has provided shared funding for network connections between the United States and other parts of the world, and NSF personnel have taken a leadership role in representing the United States in international forums. The U.S. Department of Energy (DOE) and the National Aeronautics and Space Administration (NASA) have also been active in internationalizing their research networks for some time, largely to enable "remote science" and international research collaborations. Much of this effort has involved subsidizing international connections to laboratories and research facilities of interest to U.S. scientists.

For example, DOE is working with European scientists in programs using telecommunications to conduct sophisticated experiments using U.S.-based science facilities. Telecommunications experts within that agency want higher transmission speeds to improve real-time response interactions involving scientists in Japan, Germany, and Russia, participants in the International Thermonuclear Experimental Reactor program. NASA seeks infrastructure improvements to collect and display in video format the enormous quantities of data being acquired by the Earth-observing satellites of other nations. The creation of a "Giant World-Wide NASA" as part of the international Earth Observing System will provide scientists with greater insight into global weather trends and facilitate identification of Earth resources. Both agencies are convinced that the degree and frequency of collaboration among the participants in their international programs will increase dramatically with increased bandwidth and connectivity.[1]

In parallel with the development of national networks, a number of "volunteer" networks have developed, the most prominent of which is Fidonet. Fidonet reaches 84 countries, including many developing countries for which this is the only affordable network alternative. In this regard, Fidonet technology is used extensively by nongovernmental organizations working on economic development in Africa and Asia. Electronic mail gateways allow people in such countries as Kenya, Namibia, and Ethiopia to correspond with colleagues in other parts of the world.

The global network environment now reaches over 140 countries. A variety of technologies are used, but the common denominator shared by all is the ability to exchange electronic mail. As new countries, such as those in Eastern and Central Europe and Latin America, have connected, the first international service has been electronic mail. A universal goal has been to move to the next level, interactive services and connection to the Internet. The principal barrier to such enriched communications is the high cost of international communications.

There are several reasons for the high cost of international communications. Historically, an important factor has been the high cost of installing and maintaining the physical infrastructure coupled with the limited capacity of such facilities. But with the advent of optical transmission and the introduction of new technical innovations such as optical amplification and wave division multiplexing, capacity is increasing dramatically, and hence costs can be amortized over an enlarged user and application base.

A second reason for the high cost of international communications has been the lack of competition. Until recently, most governments exercised either direct or indirect control over national telecommunications as well as international links. While this is still the case in some parts of the world, there has been a significant increase in competition resulting from deregulation and the opening of domestic and international markets to foreign companies.

Despite the growing availability of bandwidth capacity, the cost differential between U.S. and overseas service is not something that is going to be resolved in the immediate future, although there are visible signs that change is on the way. In almost every nation of the world telecommunications services operate in a "contrived" economic system, that is, a system of cross-subsidization in which monies are collected to support a wide range of government goals. In the United States one of the goals is "universal service" or the provision of affordable telephone service for the greatest number of people. The Swiss, and many other countries throughout Western Europe, use monies collected from telecommunications services to subsidize postal and transportation services. Introducing competition into these countries, such as we are now attempting to do in the United States, does not appear at this time to be in the best interests of many of these countries and, in fact, the telephone companies are discouraging anyone from building their own networks employing leased lines. Rather, the telephone companies are promoting integrated services digital network and expensive narrowband 8.25-kbps services on the basis that raising the price for leased lines high enough will encourage everyone to move toward these services. Considerable evidence exists to question that this strategy is working.

In the case of satellite communications the lack of competition is equally discouraging, resulting in fewer choices and increased costs for major users. Satellite communication was initially introduced into Europe so that the Postal Telephony and Telegraphy organizations (PTTs) could communicate among themselves and serve as gatekeepers to the flow of information in and out of their countries. Revenues from these services have been rather substantial. Two factors, possible loss of control and income, have made the PTTs reluctant to enter into a market economy for telecommunications in which competition is encouraged.

It is important to keep in mind that while networks have developed at different paces in different countries, the phenomenon has been global in nature. While the technologies most commonly used today, the Internet, BITNET, Fidonet and UUCP protocols, were developed largely in the United States, each national network activity reflects the unique characteristics (economic, legal, regulatory) of its local environment. These national efforts should be considered as peers to the U.S. networks, and it should be understood that the "global NII" will likely not be a reflection of the U.S. NII.

IMPLEMENTATION OF INTERNATIONAL CONNECTIVITY

Achieving international connections has three components: transmission between countries (in particular, transoceanic connections), distribution (local infrastructure and access points) within countries, and bilateral or multilateral agreements on technical (e.g., addressing, routing) and policy issues (e.g., acceptable use, financing).

Intercontinental Transmission

Several important considerations can affect the international connectivity of the NII. Some are technical in nature, and the others involve cost-of-service factors that must be the result of competitive forces and cooperative efforts among nations. The availability of adequate bandwidth—that is, whether or not the United States has in place, or in the planning stages, sufficient digital connections to Europe and the Pacific Rim to meet the high-speed bandwidth requirements of the NII—is an important technical consideration. A quick review of this issue suggests that bandwidth capacity will not be a problem.

The two major technologies underpinning high-speed transmission media are satellites and fiber-optic cables. In the case of satellite technology, Intelsat remains the primary service provider for most of the nations of the world.[2] In addition, a large number of regional satellites provide a range of broadcast and point-to-point services.[3]

A recent development in the satellite arena is the expected growth in deployment of very small aperture terminal dishes (VSATs), which can facilitate the supply of connectivity to remote areas and significantly enhance local infrastructure (see "Local Infrastructure" below). PanAmSat is already being used for this purpose in Latin America, and both carriers and satellite systems vendors are expected to roll out VSAT services at very low cost. VSATs should dramatically expand Internet access worldwide. This will be accomplished as federal government regulations, policies, and attitudes regarding satellite and other communications technologies, devised and implemented in the 1950s and 1960s, are changed to reflect the vast capabilities of the technologies of the 1990s. These government changes will not end at the shores of the United States, nor will they be limited to technology considerations alone. The NII planners are committed to working with their international counterparts to promote a seamless, open structure that will be most effective when it becomes integrated into a global information infrastructure. This means addressing transborder data flow policy problems such as privacy, monitoring requirements of other governments, censorship, and the exchange of certain types of data between countries.

International connectivity is also available via cable (coaxial or fiber), and as in the national context, experience suggests that the importance of fiber-optic cable for international transmission is likely to grow. One reason for this is that cable provides for better-quality service for interactive applications.

The current large-scale capacity of fiber-optic connections between the United States and Europe (TATs-8, 9, 10, and 11) will be further enhanced in August 1996 with the availability of TATs-12 and 13. The two latter fiber-optic cables will operate at gigabit rates. Looking toward the Pacific Rim, there are in place several large-capacity fiber-optic cables (TransPac 3 and 4). These have the same capacity as the TAT-8 series of cables. TransPac 5, scheduled to be in operation by 1996, will offer gigabit rates. Submarine cables provide an attractive economic advantage for selected routes where there is a high rate of growth in demand for communications capacity.

Local Infrastructure

Although intercontinental transmission technology is rapidly advancing, there is a distinct absence of a generic infrastructure that is capable of taking advantage of this capacity in most developing nations and in some developed countries.

The disparity between U.S. and foreign data communications environments (including networks and associated hardware and software) is a source of operational frustration to businesses and, in particular, members of the research community when they seek to effect or use international connections.

Within a country, distribution—local infrastructure—is largely a matter of local policy and investment; it is the area of greatest unevenness across countries. For example, in countries such as China and Russia the infrastructure for very high speed links is almost nonexistent. A major finding of the 1990 Report of the Task Force on Telecommunications and Broadcasting in Eastern Europe, commissioned by the U.S. State Department, stated: "The years of political bias against private access and dissemination of information have left all Eastern European countries, regardless of GNP, with telecommunications and broadcast infrastructures which are among the very poorest and most antiquated in the world. Massive investment will be required to bring the telecommunications and broadcasting infrastructures up to developed country levels."[4] And, of course, whatever the situation in this part of the world, conditions are far worse in most of Africa and parts of Asia and Latin America.

Since that report was written a number of major changes have been initiated by the governments of most of these countries to improve their telecommunications infrastructures, such as the move from analog to digital circuits, the introduction of cellular radio, the passage of legislation to encourage investments by Western companies marketing products and services in telecommunications, and so on. But despite the gains being made, the finding of the State Department report remains valid; significant investments are required to bring these nations to "telco equality" with developed nations. These investments depend to a large degree on the ability of these countries to promote market economies and to demonstrate a business environment that will provide investors a reasonable risk and return on their investments. Further, the financially poor condition of the universities, a widespread situation, rules out any purchase of expensive services in the near future.

As suggested above, VSATs hold the promise of upgrading local infrastructure, sometimes providing a capability that overlays an existing wireline infrastructure, at a relatively low cost. Such overlay connectivity is the goal of some new university-local business consortia being formed in Eastern Europe, encouraged in some cases by the World Bank.

In more advanced countries, such as Japan and some in Western Europe, the problems are mostly political, not based on a lack of underlying telecommunications infrastructure. In such countries, state owned or chartered PTTs control both domestic and international communications. It can be nearly impossible to acquire a direct link into a specific site, and often use of an expensive X.25-based and PTT-operated network is mandated.

When available, obtaining leased line service may involve substantial delays and/or costs that are very high compared to U.S. costs. A paper written by two information technology specialists, part of a research team in CERN working at the world's largest particle accelerator facility, lamented the fact that reasonably priced leased circuits were not available and the situation was adversely affecting their research activities. One example of a political issue is the approach in some countries toward standards. For reasons of competitiveness or, in some cases, to avoid adopting what has been perceived as a U.S. standard (the Internet protocols), some governments and intergovernment groups have decreed that national networks must use the ISO/CCITT OSI-compliant protocol suite. However, software has not been readily available and some of the standards have been slow in coming, leading to delays in implementation.

One possibility for overcoming shortcomings in local distribution is wireless communication technology, including microwave cellular telephony. In addition, cable telephone technology might also enhance local infrastructure in some countries.

It is important to keep in mind that local infrastructure is a national issue for each of the countries involved. It is not appropriate for U.S. private or government groups to try to dictate national network approaches to other countries despite the fact that many U.S. companies are competing in foreign countries for intra- and international telecommunications business. Moreover, while policies such as those cited above have limited communication between academics and researchers, they have also had the effect of stifling the development of national industries to compete with U.S. companies in this area.

Foreign Research Networking

U.S.-based research networks have provided a vehicle for overcoming some of the foreign interconnection hurdles. Some government agencies are applying their resources and traffic to specific generic networks operated by organizations within a country, supporting a major research institute in an effort to increase Internet usage. For example, in Japan NASA is leveraging its resources to enhance the Tokyo International Science Network operated by the University of Tokyo. In Russia NASA works closely with the Space Research Institute to achieve the connectivity it requires in this country.

While technology may help to lower the costs of connectivity, the essential barriers to foreign access and enhanced foreign distribution are pricing, standards, and trade barriers. These issues are discussed below.

COST AND PRICING OF TRANSMISSION CAPACITY

Any effort to create a high-speed network to Europe and to the Pacific Rim nations must first consider who pays and how payment is to be made.

Network users in Europe spend, on the average, ten times more for circuits than similar users in the United States; and in some regions of Europe T1 service is unobtainable at any price. The disparity between the costs of international circuits initiated in the United States and those of other countries seriously constrains the cost-sharing efforts of government agencies using these circuits.

Today some government agencies pick up the costs of the long-haul transmission channels and their international partners pay for the local distribution circuits within their own countries. For example, when DOE or NASA arrange international connections, they have typically provided the international circuit in exchange for provision of distribution by the foreign partner. There is an important policy question involved here: To what extent should U.S. agencies, and therefore U.S. taxpayers, support non-U.S. networking facilities even though this support might help U.S. scientists? It is likely that there will be different answers for different countries. A related issue is the degree to which the U.S.-supported facilities carry traffic other than that dedicated to the particular application that justified the link.

STANDARDS FOR GLOBAL INTERCONNECTIVITY

The international connectivity issue is made even more complex because of the different standards in place and being planned.

Achieving consensus on standards in the current international environment has been difficult at best. Historically, many countries have emphasized and promoted conformance with the Open Systems Interconnection (OSI) protocol suite, in contrast to the Internet choice of the

TCP/IP protocol suite. One U.S marketing specialist, who works for a major interexchange carrier, refers to this environment as "religious political wars" wherein government bureaucrats decide which protocols are "politically correct." His description of the standards adoption process is shared by many government network users and U.S. telecommunications manufacturers and service providers doing business overseas. They see the standards developed by slow-moving international committees over many years as not serving the needs of the user community; yet, these governments, commissions, and large research funding organizations representing one or more countries often mandate the acceptance of these standards.

However, the market for OSI products has been weak and, in the opinion of some experts, both U.S. and foreign, the whole OSI development process has been flawed because of its cumbersome design process (design-by-committee and consensus) as opposed to a reliance on experience and market forces. In contrast, commercially successful protocols (e.g., Ethernet, Token Ring, TCP/IP, X.25) have been those that have been defined, implemented, and revised based on experience prior to being standardized.

In recognition of such problems, OSI appears to have been dealt a deathblow by the Fall 1993 announcement of the European Union through EWOS that its efforts will now focus on the Open Systems Environment (OSE) rather than OSI; OSE is meant to encompass all open system standards. To a great extent this decision only validates what has actually been happening in many countries where network builders have already begun to base their networks on TCP/IP, currently the standard of choice for multivendor data networks. Where this decision was made a number of years ago, e.g., Scandinavia, the growth of research and education networks as well as the commercial availability of network services has paralleled that in the United States. Similar rates of growth are now under way in many countries throughout the world.

EXPORT CONTROLS AND INTERNATIONAL NETWORKING

Achievement of full international connectivity hinges on reducing inequities in local distribution infrastructures. One dimension is broadening access to essential technologies—first those related to physical facilities, and then those related to information services. The access issue is one where different U.S. policy perspectives appear to have collided.

The export of certain kinds of communications equipment and systems is restricted under a program of controls intended to protect national security. Some of these controls are specific to the United States, and some have been broadly shared among industrialized nations under the

Coordinating Committee for Multilateral Export Controls (CoCom) umbrella, which is expected to be replaced by a new multilateral structure.[5] Experience in the international networking arena has raised questions about the designation of products that fall under controls, the tightness of the controls, and the degree to which U.S. vendors may be or have been disadvantaged relative to vendors of comparable products in other countries. The March 30, 1994, announcement that the United States would relax many of its controls on the export of computing and communications technology gives rise to a more positive outlook than has been evident to date.[6]

In the recent past the export of many items ranging from 9,600-kbps modems to electro-optical equipment and digital subscriber interface equipment having transmission rates in excess of 156 Mbps and 144 kbps, respectively, has been restricted. Technologies for encrypting communications in the interests of privacy or security fall under even tighter controls than transmission and switching technologies. Those tight controls continue to be in effect, although they are coming under increased scrutiny.

Inasmuch as export controls succeed in limiting the capabilities in foreign infrastructure, there is a possibility that traffic originating in the United States would have to be throttled for distribution in other countries, slowing the delivery of communications, and that certain applications (e.g., video, real-time communications) would be less feasible in other countries, militating against certain forms of research collaboration, for example. Such impacts on research collaboration have already posed problems for U.S. government agencies whose missions require increased technology transfer to other countries to ensure strong international connectivity. An example is the inability of NASA to get Cisco and Proteon routers into Russia. But perhaps even more critical today are restrictions on export of equipment employing dynamic adaptive routing, i.e., network routers. This technology, developed in the 1970s with support from the Advanced Research Projects Agency (ARPA), is essential to the operation of the global Internet. And today, U.S. companies are the major suppliers of such equipment in all parts of the world. Since the algorithms used are well known, continued restrictions on the export of U.S. manufactured equipment would lead to loss of market share by U.S. companies.

The scope of this problem is such that the Federal Networking Council, an organization of 15 government agencies, put in place a special advisory group to study the problem and to make recommendations. The FNCAC indicated in October 1993 that it "endorses full connectivity of federal (and commercial) networks to non-allied countries." It argued that security mechanisms are better placed in host systems than in the network. It noted that key packet switching and cryptographic technologies are available from non-U.S. sources. It urged reevaluation and some relaxation of controls.

Demands from research and industry for high-bandwidth communications between countries suggest that there is a need to periodically assess existing export controls and to explicitly weigh and balance policy goals for protection of national security, promoting international research collaboration, and fostering foreign market development for U.S.-based producers of communications technologies. At a minimum, the performance ceiling (e.g., the bandwidth level permitted for exports) should be frequently reevaluated to prevent perpetuation of outdated rules.

To date, U.S. superiority in networking technologies and applications has been a source of competitive advantage to U.S. businesses operating internationally; not only physical facilities but also higher-level services and applications are areas in which the United States has exhibited leadership. That leadership is threatened to the extent that U.S. vendors are hindered from selling (e.g., by long delays in the export approval process) products available from vendors elsewhere. Until the recent relaxation of controls, many U.S. companies were prohibited from selling products being marketed by Japan, Germany, and other industrialized nations competing with U.S. vendors in global markets.

Controls on exports are not the only trade problem constraining the development of worldwide information infrastructure. There are business practices and export control devices being employed by other nations that may be regarded as "unfair" by U.S. standards. These practices and procedures tend to build economic walls that favor specific technical standards, products, and architectures, oftentimes initiated to limit the sales of U.S. technology abroad. The result of this is to seriously impede the growth of U.S. export markets in information technology. In addition, there are information walls involving the unequal sharing of information and information-related material on the part of our trading partners. Since the enhancement of information infrastructure in the United States will provide new opportunities for foreign parties, benefiting other nations, questions arise as to whether better access by U.S. parties to foreign information infrastructure and resources should be arranged as a quid pro quo.

CONCLUSIONS

International connectivity must be maintained and expanded as foreign networks develop and proliferate. Beyond physical access, one or more bodies (organizations) may be needed to develop and monitor bilateral and multilateral agreements on standards, transborder problems, and transborder legalities.[7] For example, as the growth of commercial networking proceeds to expand, international agreement on address space management, use of encryption technology and methods, authentication of signatures and senders, management of rights and funds transfers, and so on is needed both to maximize connectivity and communication and to enable international electronic commerce. Agreements and mechanisms also need to be negotiated to handle security problems in the international network environment, including mechanisms for communicating and resolving problems and for dealing with perpetrators. In addition, both to assure the maximum usefulness of international connections and to support U.S. vendors, export control restrictions on the sale and deployment of U.S. infrastructure technology should be reviewed and, as appropriate, relaxed.

Many players are already addressing some of these issues (e.g., the Internet Society is addressing protocols; the State Department participates in international standards and coordinating conferences; the Department of Commerce addresses export controls through its Bureau of Export Administration and comparative national infrastructure development through its National Telecommunications and Information Administration; the National Institute of Standards and Technology (U.S. Department of Commerce) and the National Security Agency have begun to address some international security issues; and so on), but a consistent framework, coordination, and mechanisms for gaining input from the research, education, and other communities are needed, and there should be a clear demarcation of authority in these areas, so that affected communities in the United States and counterparts overseas know whom to contact for what.

NOTES

1. Personal communications: Robert Aiken, U.S. Department of Energy, and Anthony Villasenor, National Aeronautics and Space Administration.

2. Intelsat is an international consortium of countries that owns and operates 18 satellites in the Atlantic, Pacific, and Indian Oceans, with which it provides voice, data, and television services. Approximately 65 percent of all revenues are derived from voice circuits. Inmarsat is a 44-member organization of nations that purchase services from it for such applications as telex, voice, fax, and data transfer up to the T1 level. Inmarsat does not own its own satellites but leases circuits on satellites operated by Intelsat and other carriers. Growing demand for increased competition in the international satellite communications industry has resulted in recent loosening of the regulation of the U.S. satellite services market, raising questions about prospects for reductions in costs and increases in innovation, both benefits typically associated with deregulation. The FCC recently decided to provide a more open environment for the entry of new systems. Five companies have been granted authority to compete with Comsat, the U.S. authorized agent to Intelsat, in marketing and delivering international communications. This authority was granted subject to a number of limitations such as not having the right to interconnect to the public network. PanAmSat, Columbia, and Orion are some of the better known companies that are in oper-

ation or in development and that are working vigorously to eliminate the limitations currently imposed on them. Recent developments indicate that the easing of these restrictions may soon be forthcoming.

3. For example, Arabsat is a multi-application satellite that provides communications services to a consortium of Arab countries. Eastern Europe and the Russian territories are served by satellites owned by a consortium of 14 countries similar to Intelsat called Intersputnik. Recently Intelsat entered into an agreement with Russia's Informkosmos granting Intelsat the option to lease three satellites in 1994. In the Pacific Rim the Palapa satellites provide services for Indonesia, Malaysia, Singapore, Thailand, and the Philippines. Eutelsat, a cooperative of 39 European nation states having the legal status of an intergovernmental organization, has seven satellites in orbit providing leased transponders principally for television and news gathering.

4. Advisory Committee on International Communications and Information Policy. 1990. *Eastern Europe: Please Stand By.* U.S. Department of State, Washington, D.C., Spring.

5. The Coordinating Committee for Multilateral Export Controls disbanded on March 31, 1994.

6. Friedman, Thomas L. 1994. "U.S. Ending Curbs on High-Tech Gear to Cold War Foes," *New York Times*, March 31, pp. A1 and D5.

7. A new NRC study titled "Bits of Power" will examine these issues in more detail.

APPENDIX

F

Key Terms

ANS, *Advanced Networks and Services Inc.*

ANSI, *American National Standards Institute.*

ARPA, *Advanced Research Projects Agency* Part of the Department of Defense. Formed as ARPA, changed to DARPA, renamed as ARPA in 1993. It is undertaking an increasing volume and range of infrastructure-related projects.

ATM, *Asynchronous transfer mode.*

AUP, *Acceptable use policy* Statement defining what kinds of traffic are acceptable over a given network (e.g., NSFNET is intended for traffic relating to research and education functions).

Backbone, Large-capacity circuits at the heart of a network, carrying aggregated traffic over relatively long distances.

Bandwidth, A measure of information-carrying capability; the difference between the lowest and highest signal frequency, expressed in hertz (cycles per second).

B-ISDN, *Broadband integrated services digital network.*

Broadband network, A flexible, all-purpose, two-way medium that will offer the high bandwidth necessary for both conventional video and high-definition television, and for still-frame displays for information retrieval, catalog shopping, and so on.

CCITT, *Consultative Committee on International Telephony and Telegraphy.*

CIC, *Committee on Information and Communication.* Part of the National Science and Technology Council.

CICnet, *Committee on Institutional Cooperation network.*

CISE Directorate, *Computer and Information Science and Engineering Directorate* One of the major units of NSF, and the one out of which NSF networking activities are run.

CIX, *Commercial Internet Exchange* A trade association of businesses engaged in providing network services that interconnect with the Internet and/or provide similar services (Alternet, PSI, Sprint, other).

Cooperative agreement, Form of arrangement between NSF and the providers of NSFNET services. Not a standard government procurement; more flexible than a conventional contract for services.

ESnet, Wide area network operated by the Department of Energy in support of its research activities.

ETC, *Entertainment, telephone, and cable television* industry complex.

FDDI, *Fiber Distributed Data Interface.*

FIX, *Federal Internet Exchange* A point of interconnection and data exchange among federal agency backbone networks. There are two FIXes, one on each coast.

FTP, *file transfer protocol* Supports file exchange over the Internet.

Full-duplex service, Simultaneous two-way communication, as over the telephone.

Gigabit Testbed program, A government-industry-academic program for developing and demonstrating high-speed network technologies.

HPCA, *High-Performance Computing Act* 1991 legislation introduced by then-senator Albert Gore to expand and extend on the HPCC (and within it the NREN) vision developed within the Executive Branch.

HPCC, *High Performance Computing and Communications* initiative.

HPCCIT, *High Performance Computing and Communications Information Technology* Subcommittee, under the Office of Science and Technology Policy.

IETF, *Internet Engineering Task Force.*

IITA, *Information Infrastructure Technology and Applications* The new fifth component of the High Performance Computing and Communications initiative.

IITF, *Information Infrastructure Task Force.*

ISDN, *Integrated services digital network.*

ISO, *International Organization for Standardization.*

ITU, *International Telecommunications Union.*

LANs, *Local area networks.*

Last mile, Popular term for the ultimate segment of the connection between a communication provider and the customer (usually residential but could also be commercial), originating with consideration of the connection between a telephone company central office and the customer premises. Telephone companies typically call that connection the "subscriber loop"; cable television or radio/wireless connections may bear different labels, given their origins as something other than the twisted copper wire pair loop historically typical in telephony. "Access circuit" is another label for the generic concept of connection between a customer or user and the main part of a service provider's network.

LATA, *Local access and transport area* A geographic region ranging from a metropolitan area to a state, created with divestiture of AT&T and used to define service areas for regulated versus unregulated services (e.g., intra-LATA local services versus inter-LATA long-haul services).

MANs, *Metropolitan area networks.*

MPEG, *Motion Picture Experts Group.*

NAPs, *network access points* Elements within the new NSF network architecture for interconnection between the very high speed backbone network service (vBNS) and other network service providers (regional and other mid-level networks, commercial services, and so on).

NII, *National information infrastructure.*

NREN, *National Research and Education Network* A component of the High Performance Computing and Communications initiative.

NSFNET, *National Science Foundation network.*

NSI, *NASA Science Internet* Operated by the National Aeronautics and Space Administration in support of its research activities.

NSTC, *National Science and Technology Council.*

NTSC standard, *National Television Systems Committee* standard for analog transmission of television.

OCLC, *Online Computer Library Center.*

ODN, *Open Data Network.*

OSI, *Open Systems Interconnection* Protocol suite with seven layers promulgated by the International Organization for Standardization.

OSTP, *Office of Science and Technology Policy.*

QOS, *Quality of service.*

SONET, *Synchronous optical network* An international standard for transmitting information over fiber, specifying standardized optical-signal formats and interfaces and integrated into the ATM packet-switching system.

TCP/IP, *Transmission Control Protocol/Internet Protocol.*

T1 service, 1.5 megabits per second.

T3 service, 45 megabits per second.

UDP, *User Datagram Protocol.*

vBNS, *Very high speed backbone network service* The small backbone network NSF plans to introduce in 1994 to replace the larger NSFNET backbone service.

WAIS, *Wide Area Information Service.*

WANs, *Wide area networks.*

WWW, *World-Wide Web.*

X.25, A protocol for data transport developed by the International Organization for Standardization as part of the OSI suite.

Index

A

Acceptable use policy (AUP), 159-60, 282
Access circuits
  bearer service (ODN) and, 88
  bidirectionality as critical factor, 86-87, 89
  cost considerations, 179, 187
  definition, 284
  network sharing and, 181
  recommendation regarding, 91
  research needed, 98-99
Access costs, 186-88, 201n
Access links. *See* Access circuits; Last-mile connections.
Accessibility issues
  balancing conflicting needs, 150, 152-53, 166n
  bandwidth requirements, 65-67, 151
  basic services, 151
  censorship, 158-60
  controlling access, 156-58, 169n
  cost reductions, 37
  costs as key factor, 123, 132, 142
  costs of infrastructure, 176-83
  education (K-12), 132
  equity of access, 149-53
  government-generated information, 154-56, 168n
  international concerns, 272-77
  Internet as access model, 30-31
  libraries, 137-40
  navigation and filtering tools, 99-100
  network access points (NAPs), 28
  payment and pricing, 152-53, 167n, 183-98
  privacy concerns, 156-58
  research community concerns, 118
  residential service, 152, 180-81, 200n
  satellite dishes, 273
  telephone service as pricing model, 37
  telephone service example, 149-50
  types of access, 40
  ubiquity of access, 152, 167n
  user charges, 8-10
Accounting. *See* Financial issues.
Acronyms identified, 282-85
Addresses for Internet sites, 75, 246-47, 250
Addressing and naming issues, 74-76, 94-95
Administration support, 206
  goals articulating, 32, 33
  limited technical scope of, 13
Applications level (ODN), 5, 49
  access circuit bidirectionality and, 87

basic services, 59-63
diagram, 53
information access services, 61-63
interconnection via, 246
research and development needs, 103-5
Archie (file access service), 61
Architecture of networks. *See* Open Data Network (ODN) architecture.
Archives as information providers, 143n
ARPA (Advanced Research Projects Agency), 103, 104, 109n, 111n, 216, 237-39
ARPANET development, 237-39
Arts and humanities research, 114, 118-19, 144n
Asynchronous transfer mode (ATM) systems, 12, 221
B-ISDN vs, 58
bearer service (ODN) and, 56-58, 108nn
definition, 282
federal policy and, 221, 223, 232n
standards-setting groups, 73
ATM systems, *See* Asynchronous transfer mode systems.
Audio servers, 60
AUP (acceptable use policy), 159-60, 282
modem limitations, 37
reserved bandwidth service, 66-67, 108n
research requirements, 117-18
BARRnet regional network, 266
map, 24
Bearer service level (ODN), 5, 47-48
ATM and, 56-58, 108nn
access circuits bidirectionality, 86-87, 89, 110nn
centrality of, 51-52, 93-94
characteristics of, 52-55
diagram, 53
integration options, 88-89
quality of service, 65-67, 93, 95
research needs, 93-95
Best-effort service, 65-66, 179, 181, 189-91
BESTNET (educational consortium), 120
Bidirectional communications paths, 86-87, 89, 110nn
Bilateral issues. *See* International issues.
BITNET, 238, 269
Branscomb report (network access concerns), 119, 144n
Browsing network resources, 60-63, 99-100
Business issues. *See* Commercialization issues.

B

B-ISDN (broadband integrated services digital network), 56-58, 108nn, 282
Backbone network services. *See* NSFNET; T3 network service; vBNS.
Bandwidth
accessibility requirements and, 151
allocation of, 65-67, 108n
data path symmetry, 87, 110n
definition, 282
education (K-12) requirements, 126
limitation as network measure, 52

C

Cable television (CATV) issues
access circuits bidirectionality, 86-87, 89
cable links, 19
high-definition TV (HDTV), 41-42n, 85
interactive possibilities, 31
pricing of services, 182, 200n
standards, 85-86, 110n
technology upgrades, 37
video services delivery, 88-89
California as networking leader, 121, 266

CalREN educational network (proposed), 266
CATV. *See* Cable television issues.
CCIRN (Coordinating Committee for Intercontinental Research Networks), 270
Censorship issues, 156-60
Centralization issues
  Internet administration, 247-49
  user support services, 262-64
CERFnet regional network, 266
  map, 24
CICnet regional network
  libraries and, 135
  map, 24
CISE (Computer and Information Science and Engineering) Directorate, 283
CIX (Commercial Internet Exchange), 222, 249
Clinton-Gore administration's role, 32, 33, 205-6
Closed networks
  international concerns, 277-79
  open networks and, 4, 44, 47
  security of, 45
CoCom (Coordinating Committee for Multilateral Export Controls), 278
Code of Fair Information Practices (updates), 254, 258-61
Collaborative computing, 115
Collaborative work via networks
  education, 125-26
  sciences, 115-16, 118
Collaboratory, 115-16
Commerce Department and networks, 210, 280
Commercial Internet Exchange (CIX), 222, 249
Commercialization issues. *See also* Financial issues; Payment and pricing issues; Publishing industry issues.
  access circuits bidirectionality, 86-87
  access restrictions, 119
  administration views, 33
  censorship, 158-60
  costs and scalability, 180-81, 200n
  customer support services, 263-64
  funding partnerships, 174-75
  government-generated information, 154-56, 168n
  intellectual property protection, 160-65, 170nn
  interconnection services, 176
  Internet oversight, 221-23
  last-mile considerations, 36-38, 45, 180-81
  needs of markets vs communities, 150, 152-53, 166n
  NSFNET, 242
  open system support, 90-91
  payment for services and access, 152-53, 167n
  privacy concerns, 156-58, 169n
  public vs private interests, 90-91, 211-13, 232n
  residential connections, 36-38, 45, 180-81
  royalties protection, 163, 164
  service availability, 150
  standards setting, 6, 218-20, 71-72
Communications infrastructure. *See* Information infrastructure; Networks; NII.
Compliance issues, 6, 67-70
Computer networks. *See* Internet; Networks; NII.
Computer Professionals for Social Responsibility (CPSR) principles, 257
Confidentiality issues. *See* Privacy issues; Security issues.
Congestion problems, 96-97, 189-90, 201n
Connectivity issues
  access circuits, 86-89
  administration's goals, 32, 33
  circuits (links) in networks, 19
  connectivity levels, 245
  costs as key factor, 123, 132, 142
  costs of infrastructure, 176-83
  entertainment industry influences, 31-32

international concerns, 272-77
Internet access, 245-49
Internet connectivity, 30
last-mile connections, 36-38, 180-81, 284-85
links (circuits) in networks, 19
payment and pricing, 183-98
proposals, 28-30
residential service, 180-81, 200n
satellite links, 273, 280-81n
standards setting, 71-72
telephone link limitations, 36-37
Coordinating Committee for Multilateral Export Controls (CoCom), 278
Coordinating Committee for Intercontinental Research Networks (CCIRN), 270
Copyright protection, 161, 162-63
copyright management (academic), 164
open issues, 100
technological means, 104-5
CoRen (Corporation for Regional and Enterprise Networking), 267
Cost recovery issues. *See* Financial issues; Payment and pricing issues; User charges.
CSNET, 238, 269
CSUnet, 266
Customer support services, 263-64

D

DARPA. *See* ARPA.
DARTnet (ARPA), 241
Data integrity, 80, 81-83, 161
Data path symmetry, 87, 110n
Database browsing services, 60, 61-63
Database sharing in education, 121
Decentralization issues, 45
Internet experience, 247-51
network control functions, 96
routing, 94
security concerns, 101-2
user support services, 262-64
DECnet networks, 238
Defense Department (DOD) and networks, 72, 238, 109n. *See also* ARPA; ARPANET.
Department of . . . *See* Commerce . . . ; Defense . . . ; Education . . . ; Energy Department and networks.
Digital communications infrastructure. *See* Information infrastructure; Networks; NII.
Digital libraries, 132, 136-40, 145-46n
research on, 104, 111n
Digital object storage services, 60
Distribution issues (international concerns), 274-76
DOD. *See* Defense Department and networks.
DOE networks. *See* Energy Department and networks.

E

Economic issues. *See* Federal funding issues; Financial issues.
Education Department (ED) and networks
Internet support, 242
leadership role in K-12 issues, 15, 209-10
Title II-D program, 242
Education issues (higher education), 119-22
copyright management, 164
California networking examples, 121, 266
database sharing, 119, 144n
financial considerations, 7-10
instructional concerns, 119, 144n
library projects, 141
network usage examples, 120-21
privacy requirements, 157
Education issues (K-12), 122-32
censorship, 158-60
commercial support, 173
costs and scalability, 180-81, 200n
costs as key factor, 123, 142-43, 173
equity in financing, 196-97

equity of access, 149-53
federal leadership required, 11-12, 15
federal policy role, 131-32, 209-10
financial considerations, 7-10
HPCC assistance, 242
infrastructure deficiencies, 122-24
Internet access improvements, 242
learning resources, 126
network usage examples, 125, 127-30
personnel training deficiencies, 124
privacy requirements, 157
teachers' concerns. 124-25, 127-30
Education issues (lifelong education), 133, 258
Electronic archiving, 114, 141
Electronic Frontier Foundation (EFF) principles, 256
Electronic mail. *See* E-mail.
Electronic money, 55, 58
Electronic publishing. *See* Publishing issues.
E-mail
as applications-level service, 59
educational (K-12) uses, 127-30
Internet and, 243
research uses, 114, 118
Encryption issues, 82-83, 101
Energy Department (DOE) and networks, 116, 231-32n, 238
Entertainment industry issues
financing of NII by, 31
size of customer base, 31
Equity of access, 149-53
ESnet (DOE network), 283
Ethical issues, 165-66. *See also* Accessibility issues; Legal issues.
balancing conflicting needs, 150, 152-53, 166n
intellectual property protection, 160-65, 170nn
policy guidelines, 166
principles of information usage, 254-61
privacy concerns, 156-58, 169n
Exchange points on Internet (FIXes, CIX), 249
Export controls and international networks, 277-79

F

Fair information practices, 258-61
Fairness issues, 258. *See also* Accessibility issues; Connectivity issues; Social issues.
Fax services, 59
Federal Communications Commission (FCC), 231n
Federal funding issues, 172-76
basis for federal support, 9
Internet support as model, 23
limitations inherent in, 10-11, 227
recommendation for, 14-15
Federal Internet Exchanges (FIXes), 249
Federal Internetworking Requirements Panel (FIRP), 215
Federal Networking Council (FNC), 230n, 251, 278
Federal policy issues, 204-33
acceptable use policy, 159-60
access circuit bidirectionality, 89-90
access circuits reengineering, 86-91
administration goals and roles, 33, 205-6
agencies involved in, 280
coordination and management role, 213-16
education (K-12) concerns, 131-32, 209-10
export controls, 277-79
government-generated information, 154-56, 168n
infrastructure deployment, 14, 223-26
infrastructure development, 91-102, 207-8, 217-28
international concerns, 277-79
Internet coordination, 250-51
Internet oversight, 221-23
Internet support as model, 23
leadership roles, 10-11, 18, 205-10, 216-17

libraries, 135
lobbying concerns, 212, 230n
ODN as technical framework, 13, 214-15, 217
OSI support, 219
procurement role, 220-21
public vs private interests, 90-91, 211-13, 233n
recommendations, 13-16, 210, 216
research and development, 91-105, 207, 217-21, 223-28
roles defined, 205
standards, 70-74, 218-20
technical scope, limitations, 13, 214-15, 217
technology role, 207, 217-28
Fiber-optic links
compliance issues, 69
federal support of, 12
international connectivity, 273-74
Internet and, 249
networks changed by, 19-20
synchronous optical network (SONET), 284
Fidonet and international networking, 271
File transfer protocol (FTP) upgrades, 61
Financial issues, 172-202
access links, 181
access circuits, 87, 181
connectivity, 176-83, 199n
cost reduction and new technology, 37
costs as key factor, 142-43
costs of access, 152, 167n
costs of infrastructure, 176-83, 199n
database sharing, 121
education (K-12) concerns, 132
entertainment industry role, 31-32
export controls, 277-79
federal funding, 9-11, 14-15, 23, 172-76, 227
fixed costs, 177-78
funding partnerships, 174-75
geographical cost differences, 180-81
interconnect charges, 183
international concerns, 271-72, 276
last-mile connections, 36-38, 181
link technology and pricing, 37
payment and pricing, 183-99. *See also* Payment and pricing issues.
public financing needed, 9
resource sharing, 178-79, 181
scalability, 180-81, 200n
subsidies, 186, 193-95
transitional financing, 186
ubiquity of access, 152, 167n
user charges, 8-10
Financial transaction services, 60
FIRP (Federal Internetworking Requirements Panel), 215
First Amendment protections, 156-60
Fixed costs of networks, 177-78
FIXes (Federal Internet Exchanges), 249
Flat-fee pricing, 183, 191-93, 200n
FNC (Federal Networking Council), 230n, 251, 278
FTP (file transfer protocol), 283
upgrades, 61
Funding issues. *See* Financial issues.

G

Gateways
applications-level connections via, 246
in networks (figure), 22
Gigabit network testbeds, 21, 27, 283
Global issues, 269-82. *See also* International issues.
Gopher (information browsing service), 61, 64
Gore, Albert (Vice President), 33, 206
Government issues. *See* Federal policy issues; Federal funding issues

H

HDTV (high-definition TV), 41-42n, 85-86
HEPnet, 238

High-Performance Computing Act of 1991 (PL 102-194), 135, 189
High Performance Computing and Communications (HPCC) initiative
education (K-12) and, 122
expansion of, 206, 242, 242
humanities research, 122
IITA as component, 29, 29, 242
NREN program and, 240-41
subcommittee (HPCCIT) for, 14
technical roots of NII, 13
Higher-level network services. *See* Applications level (ODN).
Hospitals and health care issues
commercial support, 173
online services potential, 145n
privacy requirements, 157
HPCC. *See* High Performance Computing and Communications initiative.
Humanities and arts research, 114, 118-19, 144n
Hypertext, 30
document definition by, 164-65
information access using, 61-62

I

IEPG (Internet Engineering Planning Group), 270
IETF (Internet Engineering Task Force), 219, 270
IITA (Information Infrastructure Technology and Applications) program, 29, 122, 206, 242
IITF (Information Infrastructure Task Force), 206, 213-15, 217
Information infrastructure. *See also* NII; ODN.
costs, 177-83
description, 21
federal role, 217-28
networks compared to, 21-22
Information retrieval services, 60, 64, 65
Information retrieval tools, 99-100
Information Superhighway. *See* National Information Infrastructure (NII).
Intellectual property protection, 16-65, 170nn
academic publishing, 162-63, 164
commercial publishing, 163-64
copyrights, 161, 163-65
open issues, 100
principles ensuring, 5, 254-61
research needed, 100, 104
testbeds for, 104
Intelsat (satellite transmission provider), 273, 280-81nn
Interactive video as instructional tool, 120-21
Interconnection costs, 182-83
International issues, 269-81
connectivity concerns, 272-79
cost issues, 271-72
distribution concerns, 274-76
export controls, 277-79
federal agency roles, 270
fiber-optics connections, 273
networks development, 269-70
ODN characteristics and, 45, 46
research networks, 275-76
satellite connections, 273, 274-75, 280-81nn
transmission concerns, 272-74, 276
volunteer networks, 271
International Organization for Standardization. *See* OSI protocols.
Internet
accessing, 245-49
addresses for, 75, 246-47, 250
administration characteristics, 250-51
censorship issues, 159
commercialization trends, 103
current status, 243
decentralization issues, 247-51
description, 21, 243-49
development of, 22-23, 237-39
diversification of (figure), 26
educational (K-12) uses, 11, 127-30

entertainment user base compared, 31
federal policy issues, 218-22
financial considerations, 8-9
funding partnerships, 174-75
growth (figure), 248
higher education usage, 121
interconnections, 249
international aspects of, 247-48, 270
journal distribution via, 163
library access, 137-40
networks constituting, 246-47
NII compared to, 18
NSF expansion of, 23, 238-40
NSFNET as backbone, 21, 27, 239-40
oversight of, 221-23, 250-52
privacy issues, 254
protocols, 276-77. *See also* IP; TCP.
research activities via, 113-19
routers, 247
services (figure), 244
technologies comprised by, 246-47
upgrades to backbone service, 21, 27
user charges, 251
value of, 112
Internet Architecture Board (IAB), 250
Internet Assigned Number Authority, 250
Internet Engineering Planning Group (IEPG), 270
Internet Engineering Task Force (IETF), 219, 270
Internet protocols
international issues, 276-77
OSI protocols vs, 276-77
Internet Society, 250-51
international role of, 270-80
standards setting, 219
user support via, 263
IP (Internet Protocol). *See also* TCP; TCP/IP.
ATM and, 57
bearer service of ODN and, 51-54, 93-94, 107-8nn
changes to, 50-51
connectivity and, 245
networks using (figure), 248
service features, 54
ISDN (integrated services digital network). *See* B-ISDN.
ISO (International Organization for Standardization). *See* OSI protocols.

J

JVNCnet regional network (map), 24

L

LANs. *See* Local area networks.
Last-mile connections. *See also* Access circuits.
definition, 285
economics, 36-38
residential services and, 45, 180-81, 200n
technology and, 36-37
Legal issues
censorship, 158-60
intellectual property protection, 160-65, 170nn
liability concerns, 170n
libraries, 138
personal information usage, 261
privacy concerns, 156-58, 169n
Liability concerns, 170n
Library issues, 133-41
accessibility concerns, 137-40
commercial support, 173
costs and scalability, 180-81, 200n
costs as key factor, 142-43, 146n
digital libraries, 104, 111n, 132, 136-40, 145-46n
equity in financing, 196-97
equity of access, 149-53
government-generated information, 154-56
interlibrary loan system, 146n
privacy requirements, 157
public library characteristics, 134-37, 146n
Local area networks (LANs)
description, 19
Internet access via, 245

as internetwork element, 22
ODN protocol stack and, 5
LosNettos regional network, 266

M

M-bone (multicast backbone), 103, 110n
Measuring and monitoring of networks, 77-78, 98
Metropolitan area networks (MANs), 22
MichNet, 122
Microwave links, 19
Middleware level (ODN) services, 5, 49, 55, 58
diagram, 53
research and development needs, 99-102, 103-4
MIDnet regional network, map, 24
Mobile access technology, federal support of, 12
Mobility of computing, 76-77, 97
Modem bandwidth limitations, 37
Monitoring and measuring networks, 77-78, 98
Mosaic hypertext interface, 30, 62, 120, 165
MPEG II (Motion Picture Experts Group) standard, 85
Multicast backbone (virtual network), 103, 110n
Multicast capabilities, 95, 97
Multilateral issues. *See* International issues.
Multimedia
money spent on, 32
products, 32
teleconferencing, 243

N

Naming and addressing issues, 74-76, 94-95
NAPs. *See* Network access points.
NASA (National Aeronautics and Space Administration)
international research networks, 270-71, 276, 278
networking support, 231-32n, 241
National Aeronautics and Space Administration (NASA). *See* NASA
National Information Infrastructure (NII), 1-13. *See also separate issues entries*: Accessibility, Cable television, Connectivity, Commercialization, Education, Entertainment, Ethical, Federal, International, Library, Publishing, Research, Telephone industry.
architecture. *See* Open Data Network (ODN) architecture.
benefits, 36-8
commercial aspects of, 3
compliance issues, 6, 67-70
development, recommendations for, 12-16, 209, 210, 214, 215, 216-17, 222, 225, 227
development scenarios, 32-33
educational (K-12) policy, 209-10
educational (K-12) possibilities, 11-12
entertainment industry vision of, 31-32
federal roles in, 10-16, 204-8
integrated vision of, 34-38
international issues, 7
Internet as basis for, 2-3
Internet compared to, 18
Internet vision of, 30-31
last-mile economies and, 36-38
ODN central to, 2-16, 218-20
openness required for, 3-4
principles for using, 245-61
research requirements, 15-16, 226-28
scalability issues, 74-78
standards issues, 6, 70-74, 218-20
unified NII, need for, 90-91
National Institute of Standards and Technology (NIST), 219, 280
National Oceanic and Atmospheric Administration. *See* NOAA
National Research and Education

Network program. *See* NREN program.
National Science Foundation (NSF). *See also* NSFNET.
CISE Directorate, 283
commercial developments from, 247
digital library research, 104, 111n
educational (K-12) support, 118, 122
international network support, 270
Internet expansion by, 23, 238-40
NII, expansion of role in, 216
regional networks study, 266
supercomputer interconnections, 28
National Security Agency (NSA), networking role, 280
Navigation and filtering tools, 99-100
NEARnet regional network (map), 24
Network access points (NAPs), 222, 285
diagram showing, 29
NSFNET expansions and, 28
Network architecture, 4. *See also* Open Data Network (ODN) architecture.
Network Communication Protocol (TCP precursor), 237
Network Information Center (NIC), 262
Network services. *See also* vBNS.
flexibility required of, 45-46
illustration, 34
Networks, 18-19. *See also* Internet; NII; NSFNET.
broadband, 20
costs of infrastructure, 176-83
elements of (figure), 20
federal expansion of, 23
information infrastructure compared to, 21-22
integrated (figure), 34
NII compliance, 6
principles guiding usage, 254-61
research requirements, 15
as shared resources, 178-79, 181
speeds for end users, 21
technology changes in, 26-27, 223-28
transitions under way in, 27
U.S. lead in, 12
user support services, 262-64
Newsgroups on Internet, 30
NIC (Network Information Center), 262
NII, 1-42. *See also* National Information Infrastructure.
NIST (National Institute of Standards and Technology), 219
NOAA (National Oceanic and Atmospheric Administration)
international research networks, 270-71
network use and support, 117, 241-42
North Carolina, networking in, 265, 268n
Northwestnet regional network (map), 24
NREN (National Research and Education Network) program
agencies involved in, 230n, 241-42, 241-42
development of, 29, 240-41
education (K-12) focus of, 119
federal funding, 174-75
federal policy and, 204-11, 228n
gigabit technology, 27
HPCC initiative and, 240
IITA and, 242
Internet support, 23, 239-42
NII and, 2
NRENAISSANCE Committee, vii-viii
members, iii
recommendations, 12-16
NSF. *See* National Science Foundation; NSFNET.
NSF Connections program, 119, 122
NSFNET backbone service
ARPANET and, 21, 239-40
cooperative agreement, 283
costs of expansion, 199
developments proposed for, 28-30
federal policy and, 221
internationalization via, 270
Internet and, 21, 239-40
map, 24

next-generation, 28-30
regional networks and, 239
status 1993 (figure), 24
traffic (1991-1994), 25
usage patterns and charges, 231-32n
vBNS and, 240
NTSC (National Television Systems Committee) standard, 35, 284
NYSERnet regional network, 239
map, 24

O

ODN. *See* Open Data Network
Office of Science and Technology Policy (OSTP), 23, 251
OMB (Office of Management and Budget), 215-251
Open Data Network (ODN), 2-4, 43-47
access circuits, 86-90, 96-97
addressing and naming, 74-76
benefits, 46-47
compliance issues, 67-70
criteria for, 44, 44
deployment issues, 7-10, 14
federal roles in, 10-16
mobility of computing, 76-77, 97
monitoring and measuring, 77-78, 98
network control functions, 96-97
objectives for, 44-46
openness defined, 3-4
organizational objectives, 44-46
quality of service, 65-67, 93, 95
recommendation for, 14
research and development, 91-104
scalability, 45, 74-78
security issues, 45, 78-84, 101-2
software development needs, 102-5
standards, 45, 70-74, 85-86
Open Data Network (ODN) architecture, 47-65
applications level, 5, 49, 59-63, 103-5
bearer service level, 5, 47-48, 51-55, 65-67, 88-89, 93-94
closed architecture vs, 4
diagram of, 53
information vs network service, 5
levels described, 47-51, 53
middleware level, 5, 48, 55, 58, 99-102
privacy protection, 5
routing issues, 94-95
security issues, 5, 78-84
summary, 4-5
transport level, 5, 48-49, 106n
Open networks, 43-47. *See also* Internet; Open Data Network (ODN).
architecture, 4, 47-65
closed networks and, 44, 47
commercial support, 90-91
description, 43
export controls and, 277-79
international concerns, 277-79
Internet example, 4, 43-44
openness defined, 3-4
security and, 45
Open Systems Environment (OSE), 277
Open Systems Interconnection protocol. *See* OSI protocols.
OSE (Open Systems Environment), 277
OSI (Open Systems Interconnection) protocols
definition, 284
federal support, 219
OSE as alternative to, 277
OSTP (Office of Science and Technology Policy), 23, 251

P

Packet-level connectivity, 245-46
Payment and pricing issues, 183-98, 201-2nn
access costs, 186-88, 201n
education concerns, 196-97
equity concerns, 196-97
flat-fee pricing, 183, 191-93, 202n
government-generated information, 154-56, 168n
interconnect charges, 183

link technology and pricing, 37
local infrastructure costs, 186-88, 201n
open issues, 100
prices of network expansion, 199n
resource and cost matching, 189-90
royalties protection, 163, 164
subsidies, 193-95
times of usage, 189-90
traffic congestion, 183, 189-90
transitional financing, 186
usage-based pricing, 189-91
Personal information, protection of, 258-61. *See also* Privacy issues.
PHONENET, 238
PREPnet regional network (map), 24
Pricing issues. *See* Financial issues; Payment and pricing issues.
Privacy issues, 5, 156-58. *See also* Security issues.
individuals' responsibilities, 169n
principles for information usage, 254-61
specific privacy needs, 157
Property rights. *See* Intellectual property protection.
Proprietary network architecture, 4
Protocols, 19. *See also* OSI; IP; TCP; TCP/IP.
PTTs (Postal Telephony and Telegraphy organizations), 272
Public interest networking. *See* Education issues; Hospital and health care issues; Library issues; Research issues.
Public library issues. *See* Library issues.
Publishing issues
archiving electronically, 114
censorship, 158-60
challenges of networks, 53-54, 61-62, 153-54, 167-68n
copyright management, 164
copyright protection, 100, 104, 161, 162-63
copyright protection technology, 104-5
digital libraries, 136-40
document definition control, 164-65
educational publishing concerns, 126
electronic distribution, 141, 153, 168n
electronic text experiments, 141
filtering and navigating online information, 99-100
hypertext, 30, 61-62, 164-65
information overload management, 99-100
information retrieval services, 61-63
intellectual property protection, 100, 104, 160-65, 170nn
libraries, 138-39, 146n
multimedia formatting needs, 105
navigating and filtering online information, 99-100
needs of markets vs communities, 150, 152-53, 166n
on-line projects, 141
open marketplace of ideas of WWW, 61-66
royalties protection, 100, 163, 164
technological changes, 153-54, 167-68n
user interface requirements, 99-100

Q

Quality of service issues, 65-67, 93, 95

R

Radio links, 19
Red Sage project (on-line journals), 163
Regional networks, 266-68
map, 24
NSF Connections program and, 238
usage types, 266
Remote access services, 59-60
Research and development for NII and ODN needs, 91-104, 110n, 223-228
Research community issues, 113-19.

See also Science and technology research issues.
access concerns, 119
arts and humanities, 118-19
collaborative computing, 115-16
costs and scalability, 180-81, 200n
costs as key factor, 142-43
equity in financing, 196-97
equity of access, 149-53
financial considerations, 7
humanities and arts, 118-19
international concerns, 275-76
Internet influenced by, 238-41
networking applications, 114
networks' influence on research, 9
privacy requirements, 157
scientific research, 113-18
Research issues. *See* Research community issues; Science and technology research issues.
Reserved bandwidth service, 65-67, 108n
Residential services connections, 180-81, 200n. *See also* Last-mile connections.
Routers, 19, 247
Routing network traffic, 94-95, 110n

S

Satellite links
international connectivity, 273, 280-81n
Scalability of ODN, 45-74-78
Scholarship issues. *See* Education issues (higher education); Publishing issues; Research community issues.
Schools and NII. *See* Education issues.
Science and technology research issues, 113-19
accessibility, 119
biomedical, 114-15
earth sciences, 113-14, 117, 241-42, 270-71
interdisciplinary possibilities, 115
Internet development role, 238-41
networks' influence on research, 9
physical sciences, 116, 117-18
space sciences, 241, 270-71, 276, 278
visualization applications, 117
Security issues, 78-84
closed systems, 45
data integrity, 80, 81-83
encryption, 82-83, 101
information security, 80, 81-83, 258-61
international concerns, 277-79
open systems, 45
principles guiding, 5, 254-61
research needs, 101-2
technological limits, 84
user controls, 83
user security, 254-61
wireless transmission, 109n
Sesquinet regional network, 239
map, 24
Social issues. *See also* Accessibility issues; Education issues; Ethical issues; Legal issues.
administration goals, 32, 33
balancing conflicting needs, 150, 152-53, 166n
educational concerns, 119-33
federal policy and social needs, 214
lifelong education, 133
newsgroups, 30
public vs private interests, 90-91, 211-13, 232n
Software development research needs, 102-5
SONET (synchronous optical network) standard, 223, 232n, 284
SPAN (NASA network), 238
Standards issues, 70-74
connectivity and, 71-72
consumer-oriented standards, 220, 231n
federal policy regarding, 218-20
international concerns, 276-77
NII compliance, 6-7, 67-70
OSI protocols, 70, 109n, 219, 276-77
risks to, 71-72
TCP/IP, 109n

State networks, 266-68
Subscriber loop (last-mile connection, telephones), 285
Subsidies for user charges, 193-95
SURAnet regional network, 239
  map, 24
Switching technology, analog vs digital, 35
Symmetry of data paths, 87, 110n
Synchronous optical network (SONET) standard, 284

T

T1 network service (1.5 Mbps), 21, 27, 286.
T3 network service (45 Mbps)
  definition, 286
  map, 24
  upgrade costs, 199
  upgrade from T1, 21, 27, 239
TCP (Transport Control Protocol)
  early development, 241
  multicast limitations, 96
  ODN transport level and, 5, 51
TCP/IP protocols, 109n. *See also* IP; TCP.
Teachers (K-12) and networks, 124-30
Telecommunications infrastructure. *See* Information infrastructure; National Information Infrastructure (NII); Networks.
Telecommunications Policy Roundtable principles, 255
Teleconferencing
  home education possibilities, 133
  multimedia, 243
Telephone industry issues
  access circuits bidirectionality, 86-87, 89
  community access support, 173, 198n
  equity of access, 149-50, 166n
  lobbying efforts, 230n
  PHONENET, 238
  technology upgrades, 37
  video possibilities, 31
Television (cable). *See* Cable television issues.
Television issues (broadcast)
  high-definition TV (HDTV), 41-42n
  NTSC, 35, 284
  transmission standard's limitations, 35
Testbeds for network technology, 21, 27
  convergence of TV with data, 85-86
Toward a National Research Network (1988), vii, 27
Traffic congestion problems, 189-90, 201n
Transmission issues
  international concerns, 272-74, 276
  Internet vs TV broadcasting, 35-36
  satellites, 273
  TV broadcasting standards (NTSC), 35
  wireless, 108n, 109n
Transport level (ODN), 5, 48-49, 106nn,
  diagram, 53
  limitations to, 96
  protocol requirements, 96, 107-8nn,
TWBnet (ARPA), 241

U

UCNET, 266
Universal service expectation, 149-50, 166n
UNIX systems, 242, 245
Usage-based pricing, 183, 189-91
User addressing issues, 74-76, 94-95
User authentication, 83
User charges, 8-10. *See also* Payment and pricing issues.
  free information providers and, 143
  Internet model, 251
  libraries and, 136
  sources of, 251
  subsidies for, 193-95
User interface requirements, 99-100
User naming issues, 74-76, 262-63

User privacy protection. *See* Privacy issues.
User support services, 262-64
UUCP, 271

V

vBNS (very high speed backbone network service)
  access possibilities, 28
  definition, 286
  federal policy and, 221
  upgrades, 21, 27, 240
Video (interactive), 63, 120-21
Video services delivery, 63, 88-89
Videoconferencing, 145n
Virtual network (multicast backbone), 103, 110n
Volunteer networks (Fidonet), 271
VSATs (small satellite dishes), 273

W

WAIS (Wide Area Information Servers), 62-63
Westnet regional network, 239
  map, 24
White-pages user search services, 262
Wide Area Information Servers (WAIS), 62-63
Wide area networks (WANs)
  description, 19
  as internetwork element, 22
  ODN protocol stack and, 5
Wireless technology
  compliance issues, 69
  federal support of, 12
  limits of, 108n
  security problems (radio), 109n
WWW (World-Wide Web), 61-63, 64

X

X.25 protocol, 265, 275, 285